Management of Forest Climates

Management
of
Forest Climates

V.K. Singh

RANDOM PUBLICATIONS
NEW DELHI (INDIA)

Management of Forest Climates

978-93-5111-165-8

Published in 2013 in India by

RANDOM PUBLICATIONS

4376-A/4B, Gali Murari Lal, Ansari Road
New Delhi-110 002
Phone : +9111-43580356, 011-23289044
e-mail: randompublications@hotmail.com

Reprinted 2018

Type Setting by : Friends Media, Delhi-110089
Digitally Printed at : Replika Press Pvt. Ltd.

Preface

Climate change can affect forest pests and the damage they cause by: directly impacting their development, survival, reproduction and spread; altering host defences and susceptibility; and indirectly impacting the relationships between pests, their environment and other species such as natural enemies, competitors and mutualists. A deeper understanding of the complex relationships between a changing climate, forests and forest pests is vital to enable those in forest health protection and management to expect and prepare for changes in pest behaviours, outbreaks and invasions. Investigates these relationships and the implications for forest health protection and management.

Climate change and forests are intrinsically linked. On the one hand, changes in global climate are already stressing forests through higher mean annual temperatures, altered precipitation patterns and more frequent and extreme weather events. At the same time, forests and the wood they produce trap and store carbon dioxide, playing a major role in mitigating climate change. And on the flip side of the coin, when destroyed or over-harvested and burned, forests can become sources of the greenhouse gas, carbon dioxide.

The managed forest clearly shows an increase in carbon stocks due to the suppression of unsustainable harvesting of fuelwood and charcoal, of around 5 tons CO2 per hectare per year. The village forest management regime is thus sequestering a considerable amount of carbon as shown above. From the data so far available, it is not clear to what extent emissions are being reduced in addition, since the rate of depletion of forest in the unmanaged area has not yet been established. In order to make an accurate assessment of this, data over several years will be required, and any leakage from the managed area will have to be accounted for. It is the intention of this research project to continue monitoring carbon stock changes to establish annual rate of carbon loss and predict future carbon stocks. This will form the baseline scenario against which carbon benefits of the reserved forest will be compared.

The study will be useful for researchers, policy makers and practitioners.

—*Author*

Contents

1

Climate Change of Forests

The present is both a mirror of the past and a prophecy of the future. We can study forests with various objectives in mind, but to understand their present composition we must ascertain in some measure how they have come into being. Nor can we hope to foretell their future unless we have considerable comprehension of the methods — the biological mechanics-by which forests of diverse sorts have been able to persist in spite of the many geographic and often catastrophic climatic changes which have overtaken the face of the land in the more than 300 million years they are known to have existed before the advent of man. Biologically resilient as forests have been in the past, their greatest period of crisis lies ahead.

With the advent of man about a million years ago, an additional factor greatly influencing the composition and distribution of forests came into being. As a hunter, man periodically burned great areas of forest land, perhaps to drive the game to places where it might more easily be taken, or even to increase the amount of game by producing larger grazing and browsing areas; but more often through carelessness with his campfires.

With the domestication of animals and the invention of agriculture well over ten thousand years ago, his influence became greatly intensified, for man is a mobile and hungry creature ever thrusting his pastures and fields into new territory. Today, with an ever expanding population, this thrust is almost entirely into forest land. Therefore it is timely for man to pause and, in an unsentimental and objective manner, consider how forests came into being, how they developed and how forests and man fit into the general pattern of the world of which he is today a part — a world in which the future of all mankind lies.

THE CLIMATES OF THE PAST AND PRESENT

The planet on which we live is an oblate spheroid, its equatorial region being warmer than the two polar areas. This temperature gradient, in conjunction with the earth's rotation, sets up a reasonably standard pattern of atmospheric circulation. Since the surface of the earth is neither all land nor all water, but divided into land masses surrounded by even larger areas of ocean, the physical properties of water tend to decrease seasonal contrasts near continental margins relative to those in interior regions. Any enlargement of continental areas would tend to increase the seasonal contrasts, induce steeper barometric gradients and thus heighten the general storminess of their interiors.

Conversely, any decrease in the continental areas, as by a lowering of their general masses with an encroachment of oceanic waters, would tend to bring about an amelioration of climate and a general stabilization of seasonal variations. As the geological record clearly indicates, there have been times when the parts of the continental masses exposed above oceanic waters increased for a while and then shrank, with long periods of relative quiescence between, often accompanied by extensive invasions of the sea.

The periods of increase of the exposed portions of the continents were, on the whole, correlated with periods of mountain building and the development of plateaus. The geological record also indicates that these periods of continental uplift were usually accompanied by intermittent glaciation, at least in the polar regions. Volcanic activity of greater or lesser intensity, especially near the continental margins, also was characteristic of such times.

The fact that we happen to be living in one of these cycles of disturbance focuses our attention on the climatic (and consequent vegetational) features of such periods rather than on the more quiescent and far longer periods that have intervened between them. As Russell has pointed out, man during his entire history has not known what it is to live in a "normal" climate. One may also add that, during, his existence, man has not witnessed what might be termed a "normal" phytogeographic pattern.

If the surface of the earth were uniform there would be three parallel zones of higher-than-average precipitation, one in the equatorial region and the other two in the general regions of the 60° parallels of latitude. There also would be four zones of lower-than-average precipitation, two in the polar regions and two centering about 30° N and S latitude. Because of the rotation of the earth and the interaction of factors resulting from the positions of continental masses and oceans — and their unequal distribution — this theoretical pattern is greatly altered. Furthermore, as Rossby has noted, in the northern hemisphere the much greater percentage of land tends to disrupt the expected pattern more than it is disrupted in the southern hemisphere with its much larger areas of ocean.

There is a close correlation between effective precipitation and vegetation. Hence one may draw certain inferences regarding precipitation from a generalized map of the world's vegetation types. In spite of the heat, the precipitation near the equator is such that few desert areas are found there. Apart from polar wastelands, subhumid and essentially treeless regions tend to develop at the western continental margins at about latitudes 20°-30° and to trend inland in a northeastward direction in the northern hemisphere and southeastward in the southern hemisphere. Although local conditions such as mountain ranges modify these relationships, this, on the whole, is the general pattern.

The belts of heavy precipitation to be expected on a uniform-surfaced world at about 60° would also be turned northeastward on the continents of the northern hemisphere and southeastward in the southern hemisphere. In the skewing of these belts they would be depressed about 100 at the western continental margins. Their poleward trends would soon carry them into the polar subhumid areas; thus they would disappear as belts and become more or less regional coastal areas centering at about latitude 50°. Evidences of these are to be found in the northern hemisphere in the areas of heavy precipitation in North America in the Pacific Northwest and also in northwestern Europe. In the southern hemisphere only South America extends into sufficiently southerly latitudes for the southern superhumid area to be fully evident, as in southern Chile; but here no extended area is found, because of the southern Andes and the small total amount of land area available.

Reference to the vegetation map will indicate that the tropics, which we sometimes think of as characterized by heavy precipitation and an extensive development of rain forest, are by no means uniformly so. The less humid areas just north and south of the equatorial regions restrict the band of equatorial rain forest and other factors shift it and break it into segments.

Those areas, loosely referred to as "the tropics," actually are not characterized by a heavy effective precipitation and dense rain forest; the latter is a forest type of relatively limited extent. The montane rain forests sometimes developing on the windward slopes of mountains in frost-free regions are superficially similar to lowland rain forests, but the two would be distinguished in any detailed analysis of forest types. On the basis of theoretical climatology, those longer periods in the history of the world when its surface was relatively quiescent, with the exposed portions of the continents smaller than at present and also with somewhat subdued relief, were times of "normal" climate. Vegetational zonation poleward from the equator was evident, but without the sharp gradients of today. The air may have been slightly more humid than it now is over large areas, but the total rainfall probably was somewhat less.

The latter fact would tend to decrease the extent of the expected "standard" areas of superhumid climate, which were probably little better developed than they now are. Contrarily, with a more uniform air circulation, the subhumid zones entering the western margins of the continents, while in evidence, probably did not give rise to desert conditions except in the interiors of the largest of the land masses. Again on the basis of climatological considerations, it can be postulated that, during periods of world-wide continental depression and quiescence, the frost-free area of the world extended much farther poleward than at present. Theoretically the frost-free area would have reached to the neighborhood of latitudes 60° N and S; actually, judging from plant remains, it may have extended somewhat farther poleward on occasion. However, where one places the frost-free limits in relation to the 60° parallels depends to a certain extent on whether one does or does not believe in the theory of continental displacement.

Lastly, on purely theoretical grounds, the polar regions of such periods would be characterized by a cool but not cold climate. Snow probably fell in winter, but along the margin of any polar sea which existed the climate is not likely to have been much more severe than it now is in southwestern British Columbia, northeastern Nova Scotia, northern Scotland, or southern New Zealand. The abundant plant fossils of both the Arctic and Antarctic regions offer ample testimony that these theoretical speculations are not far wrong; if theory errs, the plants indicate that it may be on the side of conservatism.

The general configuration of the climatic pattern of the lands of a hemisphere during periods of general continental lowering. It may be taken as roughly diagrammatic of the combined land masses of North and South America (together with the present land of Antarctica), or else of the combined land masses of Europe, Asia and Africa; in both cases there is more land in the northern than in the southern hemisphere. During the periods represented, the climate of the world was more generally equable than it is today. With no polar ice, the land masses nearest the poles, as just noted, enjoyed reasonably temperate conditions, with the warm-temperate zone of transition to frost-free climate at about 60°.

The somewhat lower temperatures of what then were the temperate zones produced a condition sufficiently humid to limit the poleward extension of the midlatitude subhumid areas. The superhumid areas, centering at about the 50° parallels, soon dwindled in the interiors upon meeting the zone of low polar precipitation. In spite of the wide latitudinal extent of the frost-free zone, the equatorial region, on the whole, was probably but little warmer than at present. While there was a superhumid equatorial area, the general lowering of precipitation on a world-wide basis did not permit it to be very extensive. Low mountain ranges may have altered the precipitation in particular areas and brought about local or regional departures from the pattern as here outlined.

The corresponding general configuration during periods of continental uplift and disturbance. At such times the frost-free zone was greatly reduced in extent, the temperate regions were moved equatorward and great areas of what we now think of as having a typical "polar climate" developed in the Arctic and Antarctic regions, accompanied by icecaps and extensive intermittent glaciation. With these changes came a marked enlargement of the subhumid areas, originating at about 30° and a considerable increase in their aridity, especially in the continental interiors, where great wind-scoured deserts were formed. During periods of extensive continental glaciation, the lowering of the temperature along the ice fronts locally — or even regionally — reduced the extent and severity of these subhumid areas, as it did in the Great Basin area of North America during the ice maxima of the Pleistocene. The maxima of periods of disturbance, when the "polar climates" and continental glaciers descended to beyond the 40° parallel, as they did in certain regions several times during the Pleistocene, but a somewhat more equable condition such as we have today. The Permian appears to have been characterized by periods of much more severe climate than was the Pleistocene.

The Forests of the Past

The development of the vegetation types — and especially the forest types — of the past. It should be held in mind throughout that the continents, in those periods when they were of more subdued relief, nevertheless were by no means topographically featureless. The incompletely eroded bases of former high mountain chains might have lingered on, even when the next cycle of mountain building was already beginning. The face of the land was varied and its elevations, if not on the magnificent scale of the present Andes or Himalayas, still were sufficient to bring about modifications in the precipitation and local climatic pattern. And these elevations — the uplands of the past — have played an important role in the development of the world's vegetation, for it becomes evident, when we consider the details, that in great measure it was on such uplands that the new groups of plants underwent their primary evolution.

FORESTS OF THE CARBONIFEROUS

The Carboniferous saw the extensive development of lowland forest types, as we should expect of a period when the continents were generally at a low level. Again, however, we must presume a certain amount of upland, even with mountains of possibly considerable height, for there are far too many fragments of trunk and branch wood, with no associated leaves or other structures with which they can be related, to be accounted for in any other way.

Furthermore, where else but from adjacent mountains and extensive uplands could the sediments producing the abundant sandstones and shales of the time have been derived? But it is to the lowlands, with their rich fossil record, that we turn our immediate attention.

It has usually been thought that the extensive Carboniferous lowlands were characterized by extremely heavy precipitation and there is no doubt that swamps were widespread. What sort of "swamps" were they? The answer is to be found in the trees that inhabited them. To live and function actively every cell must have a relatively abundant supply of oxygen and also be able to get rid of any excess of carbon dioxide produced during certain physiological reactions. Therefore, when we examine the plants of constantly wet areas, we find that those portions where the aeration is poor develop a sort of internal tissue known as aerenchyma.

This specialized tissue consists of a series of ramifying ducts, channels, or connecting loose-celled structures whereby the living cells of the submerged portions can exchange air with the parts in normal atmosphere. There is little point in arguing, as some have done, that "plants were different then." Protoplasm was protoplasm, just as it is today and there is absolutely no evidence that it had any different requirements or tolerances. The forms of the plants have, perhaps, changed somewhat (fundamentally very little), but the basic physiology seemingly has not. Therefore, we can examine the roots of these groups of Carboniferous plants and tell with considerable precision whether or not the soil in which they grew was saturated with water.

Of the roots of Carboniferous plants that have been examined in any detail, only the following are known to have possessed aerenchyma: the equisetopsids in general (the sphenophylls and arborescent calamites), the marattiaceous tree fern *Psaronius* and perhaps a few, but by no means all, of the arborescent lepidodendrids. The great bulk of the lepidodendrids and all the other arborescent forms of the time, so far as is known, were devoid of it, indicating that they did not inhabit areas which were flooded and waterlogged for considerable periods. But how, then, can we explain the great deposits of coal in the Carboniferous, if they were not laid down in vast, reeking swamps? Here again the Coal Measure strata tell the story.

These abundant plant remains accumulated in slowly descending basins. Near the estuaries, which often extended well into the continental interiors, in actual depressions inland and along streams, the equisetopsids, *Psaronius* and a few of the many species of lepidodendrids, with their specialized root aerenchyma, were abundant. With an equable and completely frost-free climate extending at least to latitudes 60° N and S, there would be ample opportunity for rampant growth over a large portion of the land surface if soil moisture were reasonably available.

In the slowly descending basins characteristic of large areas of the land surface at various periods of the Carboniferous the soil moisture was certainly ample at all times of the year. However, this does not mean that the rainfall was necessarily excessive or the soil saturated with water, except locally. With the generally low relief of these basins would have been associated a complex pattern of saturated, poorly aerated soils, intermingled with soils that were moist but had sufficient aeration for the growth of plants without a specialized aerenchyma in their roots. So intricate was the pattern and so dense the vegetation that as the trees fell, to be preserved in the deposits, they fell in such a way as to make it seem at first glance as if they had all grown in exactly the same habitat.

The same intricate pattern in the distribution of different species is easily discerned in our modern swamp forests, where we find some species capable of persisting with their roots submerged, while others, growing so close as to have their branches touching, may be seen to occupy slight rises on which the soil, although constantly moist, is much better aerated. These latter species also are likely to be found wherever the soil moisture is abundant outside the swampy zone, sometimes even on uplands.

Although moisture was reasonably constant in the lowland soils, the humidity of the air was not excessive. Careful study reveals that the great majority of the Carboniferous plants had leaves which in some manner resisted water loss through transpiration. This applies to the arborescent lepidodendrids, the pteridosperms and the cordaites — the last a group of lowland pinopsid trees then in great abundance, which sometimes formed. dense stands. Even the microphyllous equisetopsid trees of the time, growing in their soggy environments, indicate that the atmosphere was by no means saturated with water vapour.

For our purpose, however, it is perhaps not overly important what the conditions were under which these lowland forests grew, except insofar as they shed light upon the worldwide atmospheric and climatic conditions of the time and thus give clues as to what was probably happening on the even drier uplands, where the ancestors of our contemporary trees already were undergoing their primary development.

Actually, we have little first-hand evidence regarding the upland forests of the Carboniferous, apart from what is revealed by the relatively few fragments of trunk and branch wood that floated into the lowlands, there to be preserved in the deposits.

While these give us valid hints, it is from the groups that migrated into the more equable lowlands during the severe climate of the Permian that we can glean a more satisfactory picture of the upland forests of the Carboniferous. Just as the wood fragments indicate, these migrants consisted mostly of a series of pinopsid forest types, with forerunners of the true "conifers" dominant among them. Often made up of giant trees (specimens with trunks up to six feet in diameter are known), these pinopsid forests probably had an understory of pteridospermous trees, shrubs and vines (with perhaps a few herbaceous types) and there is every likelihood that ferns of diverse sorts were abundant.

The subhumid areas near the 30° parallels probably were not devoid of vegetation. The tough-leafed pteridosperms of the time were already giving rise to the cycads and cycadeoids, both admirably fitted to less humid — even xeric — conditions. Therefore, it is probable that the less severe parts of these subhumid areas supported a variety of pteridosperms suited to such conditions. And a few scattered fossils indicate that ancestral elements of certain lines, leading to those of the modern pinopsids better suited to dry areas, already were present. Cordaitean forests dominated the moist lowlands;

ancestral "conifers" were beginning to take over the uplands. If any forests grew in the cool-temperate regions and on the slopes of any higher mountains of the Carboniferous period, their nature is largely a matter of conjecture. If, however, the requirements necessary in the present are a mirror of those of the past, there is every reason to believe that abundant forests were then present in the Arctic and Antarctic regions, unless they were limited in some manner by the lack of light during the long polar winter; certainly the temperatures were no bar to the development of dense forest stands.

If so large a number of pinopsids can today stand temperatures far below freezing, surely the large and variable group then extant could have included forms capable of persisting in the polar regions in the less rigorous temperate climates of the time.

Their apparent abundance on upland habitats at middle latitudes during the period is an almost certain indication that the pinopsids also included a wide series of types adapted to temperate climates. And since resistance to frost in plants is physiologically closely akin to resistance to drought, there is considerable reason to believe that the poleward temperate regions of the time also supported at least a scattered assortment of pteridospermous types.

Although there is little or no evidence that the vast lowland forests of the Carboniferous contributed anything to the development of modern arborescent types, they played a most important role in relation to the use and ultimate preservation of our present-day forests.

Before the widespread use of coal, wood was almost the only source of fuel. In viewing our dwindling forest resources, we may be thankful that the lowland forests of Carboniferous time, entombed in the strata of the Coal Measures, have relieved our modern forests of much of this great and potentially destructive burden.

THE GREAT CHANGE IN THE PERMIAN

The Permian opened with a climate very similar to that of the latter parts of the Carboniferous and with the same general distribution of forest types. This pattern was soon disorganized by what may have been the greatest period of glaciation that the world has ever experienced.

By comparison, it dwarfs any of the glacial cycles of the Pleistocene, which terminated the Tertiary. In the northern hemisphere evidences of Permian glaciation are known from Alaska, Massachusetts, England, France, Germany, the Ural Mountains and India. In the southern hemisphere all of what is now Australia was covered and extensive glacial deposits of the same age are reported from Madagascar, various parts of southern and southwestern Africa and South America, notably Brazil, Argentina and Bolivia. Nor is there any hint that these deposits were derived from local mountain glaciers; they were of a continental type.

The evidence of widespread glaciation during the Permian in the southern hemisphere, coupled with that of an amazing glaciation in what is now tropical

India in the northern, did as much as anything to spark the far-from-settled controversy concerning the possibility of continental "drift." If, as postulated by the proponents of the theory, these glacial deposits could have been laid down in polar regions and then later transported into what ordinarily have been thought of as the temperate and tropical regions by a lateral shifting of the continental masses, there would be no need to disturb the "normal" climatic belts of the world.

The possibility of at least a certain amount of continental displacement need not be ruled out. But, the story told by the forest types of the time indicates that a vast world-wide change in climates actually did take place: those groups which, since the Devonian, had been limited chiefly to frost-free regions and apparently fixed in their evolutionary potentialities were essentially wiped off the face of the earth.

The climatic situation during the height of the Permian glaciation. The completely frost-free zone may have dwindled to a narrow band in the equatorial regions, thus cutting down a multitude of tropical genera and species. The continental uplifts would have drained the great Carboniferous swamplands, thereby eliminating other arborescent types with which the world had been so long familiar and interior plateau deserts with severe climates would have been extensive. It was a time of crisis for all the old forms of life — animal and plant and many of them failed to meet the challenge.

Since there is little point in dwelling on what disappeared, let us turn to that which survived. Actually, the frost-free areas may have been rather more extensive than some students have supposed, for a few of the Carboniferous lepidodendrids, calamites, cordaites and pteridosperms passed through the perilous times of the Permian and lingered on into the Triassic. Furthermore, the fossils tell us that the land unoccupied by glaciers was never barren of vegetation except in the worst of the and areas and in the polar regions; there were vascular plants of one sort or another right up to the ice fronts. And the rapid development of tropical lowland forests in the Triassic also points to the presence, in the Permian, of considerable areas of at least warm-temperate climate.

We leave this most interesting and critical period in the world's vegetational history with only one last glance. Various modern types of fern had become more evident. The cycads and cycadeoids were now common and venturing into the drier habitats then widespread. The ginkgos, the true "conifers," and perhaps even the earliest of the taxads were abundant, not only in the lowlands, but also on those uplands where the climates were not too severe.

The pteridosperms, although on the wane, still displayed a wide variety of forms. Among these were groups which, for want of a better term, might be collectively called the "aberrant pteridosperms" — plants often with entire margined leaves and a reticulate venation. The male reproductive cells, or sperm, still were motile and for this reason they have traditionally been classed with the pteridosperms, although recognized as distinct orders and families. Are we here getting our first glimpse of the real ancestors of the magnoliopsids? If these considerably varied "aberrant pteridosperms" actually were an independent complex of forms out of which the magnoliopsids were derived, then the Permian marks the really important point of change in the floras of the world, a closing out of the old forms and a bringing into prominence of the basic members of all the important modern arborescent groups.

THE GREAT PINOPSID EXPERIMENT OF THE TRIASSIC AND JURASSIC

Contemporaneously with the dwindling of the Permian ice fields and the return to a more equable climate in the Triassic, something approaching an ecological vacuum came into being, for the old forms dominant during the Carboniferous had essentially disappeared. The disturbances of the Permian had given rise to a large series of upland habitats, which were exploited by the remaining plant groups of the time.

With the newly available, although still relatively small, lowland areas freed from competition by the old types, the newer types filled the many vacant ecological niches of these regions. Being in the ascendency and already the dominant upland forest type, the pinopsids began to develop an extremely divergent series of derived lowland forms.

Many of the Permian types held on into the Triassic, but in the expanding frost-free lowlands broad-leafed pinopsids began to be evident, suited to such environments. Certain modern pinopsids, such as the broad-leafed podocarps, appear to be collateral descendents from just such types. And it seems likely that a few, such as the modern *Phyllocladus,* abandoned leaves and produced broad leaflike structures by greatly flattening and expanding their lateral twigs. In the temperate lowlands further varieties of leaf forms showed themselves, some suited to humid and others to drier conditions.

In Triassic times deserts, residual from those of the Permian, or produced in the rain shadows of far-flung mountains, probably existed and the widespread occurrence of numerous cycad and cycadeoid types suggests large subhumid areas. But the giant pinopsid trunks, abundant in the Petrified Forest of Arizona and elsewhere in Triassic formations, testify to extensive forests and the opinion expressed by some that the Triassic world was largely dominated by true desert climates is probably erroneous. In the earlier stages of the Triassic the land masses were still elevated and so afforded relatively few places where adequate deposits could be formed; therefore, we know little of what must have been the highly variant forest types of the time, save what can be deduced by observing the situation during the transition to the Jurassic.

A further lowering of the land masses in the Jurassic led to considerable amelioration of the climates of the world. Since it also gave opportunity for the formation of more extensive deposits, we know considerably more of the forest types of the Jurassic than we do of those of the Triassic. The cycads and cycadeoids were still abundant, indicating the persistence of subhumid areas of considerable extent. Along with the many old and now extinct groups of pinopsids, modern genera began to appear.

About this time the ginkgos perfected a functional abscission layer in their leaf bases, which permitted their leaves, instead of hanging until tattered and unrecognizable, to be blown about as they fell off the trees each autumn and so to settle in depressions and become preserved. These characteristic and abundant leaf impressions permit us to ascertain that the group was widespread and common in the northern hemisphere, probably with a goodly number of species.

The great bulk of the fossils has led to the conclusion that the forests of the Triassic and Jurassic were dominated by pinopsid types. But let us again remember that, although the continental masses apparently were being slowly lowered, there still were broad uplands and considerable mountainous territory, whose covering vegetation is necessarily almost unknown.

As Axelrod has pointed out, the earliest materials which can be placed among the flowering plants (magnoliopsids) with any degree of reliability have been found in Triassic strata. Materials about which there can be no argument are of increasing frequency in the Jurassic.

Furthermore, by the Jurassic even these fragmentary remains indicate substantial diversity of form, with evidence that both monocotyledonous and dicotyledonous types already were in evidence. There is only one conclusion: beginning with the Triassic and more abundantly in the Jurassic, the upland and mountainous regions, as distinguished from the more widespread areas of deposition, bore an increasingly diverse assortment of surprisingly modern magnoliopsid forest types.

By the end of the Jurassic the great experiment in diversification by the pinopsids was almost over. They had filled the numerous and often special ecological niches left vacant by the passing of the lowland forest types of the Carboniferous and Permian.

They had not only achieved world-wide dominance, but had produced a diversity of kinds which they were never to display again. Of the 55 genera of pinopsids still living, 47 may be classed as relics in view of their restricted or highly disjunct present distributions, or else because their distributions today are much smaller than were those shown by their remains in Cretaceous and Tertiary deposits.

THE SORTING OF THE FLORAS IN THE CRETACEOUS

Complex and varied as they were, the pinopsids between the Permian and the Cretaceous are known from such scattered localities that it is difficult to determine and compare the distributions of the various families and genera.

It becomes obvious, however, that even before the Cretaceous there was a strong divergence between genera, some being predominately of the southern hemisphere and others characteristic of the northern.

This separation has since been greatly intensified. Today only a few pinopsid genera have occasional species that cross the equator and these are mostly along transequatorial mountain ranges, where they may be recently adventive.

The more critically we study our flowering plants (magnoliopsids), the more it becomes evident that the groups we are prone to think of as characteristic of the temperate regions actually have their more primitive living members in the frost-free parts of the world or have been patently derived from other groups essentially tropical in composition. In view of what we know of their distribution in the Jurassic, in conjunction with the foregoing situation, we are increasingly led to the conclusion that the magnoliopsids underwent their primary evolution, not in the lowlands, but on the uplands of frost-free regions.

Since there is reason to suppose that the magnoliopsids were actually present during the earlier phases of the Triassic, this primary evolution would have taken place roughly between latitudes 30° N and S, the probable limits of the frost-free conditions of the time.

Forms suited to a somewhat cooler climate might have come into being in the Jurassic, with further segregation and evolution during the Early Cretaceous of others better adapted to even more temperate conditions. In his excellent summary, based on paleontological evidence, Axelrod points out that by the Mid-Cretaceous this ancestral magnoliopsid flora had become widely deployed in the frost-free regions and had also given rise to groups characteristic of the northern and southern temperate zones. There is a suggestion that the Paleocene — the transition from the Cretaceous to the Tertiary — saw an even wider development of a characteristically pantropic flora.

Interesting as was this deployment of the magnoliopsids during the Cretaceous — a large pantropic series, with considerably different forms in the north and south temperate regions — the forests of the time were by no means lacking in pinopsids. The latter were abundant in the cooltemperate regions and some have remained there as characteristic forest types to this day.

In the frost-free and warm-temperate parts of the southern hemisphere, broad-leafed pinopsids competed successfully with the magnoliopsids, as they still do over large areas; and forms apparently better suited to cool-temperate conditions, such as then existed on what is now the Antarctic continent, were also represented. In the northern hemisphere the lowland forests of the Early Cretaceous were primarily composed of pinopsids, with a few scattered magynoliopsids, whereas by the close of the Cretaceous the arborescent magnoliopsids had increased to about 60 per cent of the total known flora, the pinopsids had dropped to about 15 per cent and the remainder were mostly ferns, cycads and highly evolved herbaceous materials, such as sedges and grasses. There is considerable evidence, however, that, except in equatorial regions, the uplands of the Late Cretaceous still contained abundant pinopsids.

In mapping forest cover, foresters, as a justifiably practical expedient, usually designate the stands on the basis of the dominant species present today. This practice may, however, at times give an incorrect picture of the natural forest cover, especially where the stands are known to be in ecological imbalance because of disturbance. For example, it is today usually considered expedient to map large areas in the southeastern United States as being dominated by "coniferous forest."

However, as every forester in the region knows, the constant problem is to keep a stand of pine on the more productive sites from going over to a mixture with hardwoods which, to him, are "weed species." Only on exceptionally poor sites do the pines maintain their dominant position without assistance. This is clear evidence that the region basically is one of a mixed forest type. Today controlled burning is being ever more widely accepted as a legitimate silvicultural method for the control of hardwoods in the extensive southern pine forests. In a similar vein, evidence is only now accumulating that the once extensive stands of white pine in the New England states were the result of recurrent hurricanes which, periodically for a very long time, have swept the area with destructive force.

Under undisturbed conditions in New England hardwood forests dominate the stands, with white pine scattered, any pure stands of it being limited to rocky outcrops and patches of extremely sterile soil. On more fertile soils in New England white pine is merely the first stage in a normal ecological succession following extensive blowdown by hurricane winds, or after clearing or fire. Here, as in the southeastern states, mixed forests of magnoliopsids and pinopsids would be the natural climatic cover.

Thus, without going into details on a world-wide basis, we may here note that mixed pinopsid-magnoliopsid forests, exhibiting an infinity of variations from region to region, but always with much the same general physiognomy, are widespread under warm- to moderate-temperate conditions if the rainfall is adequate and the areas but little disturbed. In fact, when examined objectively, the so-called "coniferous" forests of cool-temperate regions usually have admixtures of hardwood species.

As far as associations of species are concerned, these and similarly mixed forests, scattered across the temperate regions of both the northern and southern hemispheres, belong to what have been designated as various segments of the "Tertiary Forests"; but, as types of forest communities, they are more ancient and have their roots in the mixed pinopsid-magnoliopsid forests of the Cretaceous.

PRELUDE TO THE PRESENT (TERTIARY PERIOD)

By the end of the Cretaceous all the important families of magnoliopsids containing arborescent members were present and many of our modern genera were abundantly represented. The Tertiary marked the elaboration of species. In the Early Tertiary, the frost-free regions were somewhat more extensive than they had been in the latter part of the Cretaceous and offered ample opportunity for the basically warm-climate magnoliopsids to migrate. This probably explains the wide dispersal of so many tropical genera today, genera already in existence in the Tertiary. It also applies to the genera which today are widespread or exhibit disjunct distributions in the north temperate regions (Holarctica). Similar disjunct distributions of a series of families and genera of both magnoliopsids and pinopsids in the more temperate regions of the southern hemisphere, which seem never to have existed elsewhere, however, offer an incompletely resolved problem to paleogeographers.

By Mid-Tertiary time, the climatic pendulum was swinging back again. The frost-free zone was slowly shrinking, treeless areas began to appear in the polar regions and the subhumid zones exhibited considerable expansion, their inner portions becoming arid. This series of shifts culminated in the first of the Pleistocene glaciations. As is well understood, a succession of ice maxima occurred during the Pleistocene, with interglacial intervals when the climate appears to have been considerably more genial than it is today.

It has been calculated that the last ice maximum, covering perhaps as much as 12 million square miles of the earth's surface, locked up so much water from the world's oceans that their level was lowered about 300 feet.

The fact that the melting of the ice still remaining in both hemispheres would raise the surface of the world's oceans perhaps as much as 150 feet above its present level, suggests that we may be scarcely more than half way through the present glacial regression.

There is no point here in entering the lists with those currently arguing over the actual extent of disturbance of the climatic zones during the height of the Pleistocene glaciations. The shifts certainly were not so marked as during the Permian and the floras of the world appear to have recovered lost territory with considerable ease, except in certain areas.

Where there was ample room for migration, some species that appear to be closely allied to (or perhaps the same as) species in Late Tertiary strata moved southward ahead of the ice and since then have been migrating northward again. The more genetically plastic groups apparently returned after each ice maximum, with a considerable accumulation of variability leading to the production of new species.

The Forests of the Presen

We are now ready to take our theoretical representation of the climatic zones of a period of world-wide topographic and climatic disturbance and relate it to the vegetation types of the present. Instead of climatic zones, generalized vegetation types appear.

Here, again, it should be emphasized that local conditions, such as extensive mountain ranges, plateaus, or the relation of land and ocean masses, modify the details of the pattern. The map however, shows that the generalities hold.

POLAR REGIONS

The north and south polar regions retain icecaps on the lands and considerable amounts of floating ice in the adjacent seas. Although vegetation occurs on summer snow-free areas (the precipitation is low and snow cover often very thin), it is stunted and presents the general aspect of *tundra.* Locally, treeless tundra may extend below 60° latitude.

GROUPS OF VASCULAR PLANTS

Before we go further into the historical development of forests and forest types, it might be well to note briefly the various major groups of vascular plants. This is perhaps advisable because several groups which dominated the forests of the past are now almost extinct, having left behind only a few living remnants which no longer are arborescent. Were it not for their pasts they would be outside the scope of this discussion. The designations of the major groups of vascular plants may not be familiar to all. Formerly it was customary to divide the vascular plants into two groups, the *Pteridophyta* (reproducing by spores, as in the ferns, horsetails and lycopods) and the *Spermatophyta* (reproducing by seed). Advances in our knowledge have revealed that this simple division is no longer tenable.

Currently seven major groups may be recognized; these appear to have been independent as far back as we have knowledge of them in the fossil record. Fortunately, each of the major groups as now defined still contains living members. Therefore each group name has been derived from the name of a living genus, athough the extinct genera may have been far more numerous. Thus, the *Equisetopsida* are more or less like the modern horsetails *(Equisetum)* and the *Magnoliopsida* (the so-called "flowering plants") are more "*Magnolia*-like" than "fern-" or "*Pteris*-like" *(Pteropsida)* and so on. In the following text, informal usage is sometimes made of the more formal group designations.

The present discussion cannot do more than briefly mention certain matters relative to the development of these groups. Acquaint us with the names of these major groups of vascular plants, their most prominent subgroups, their known extent in geological time and their general prominence in the world of the past as revealed by the fossil record. However, at all times we must remember that the "fossil record" to a great extent is composed of members of lowland floras which were present near estuaries or in descending basins of deposition and whose remains were thereby readily entombed in silts and muds by seasonal floods, or in the acid muck of great swamps then essentially at sea level.

The variety of materials then growing on the much more extensive uplands can be conjectured only from clues in the form of logs and miscellaneous scraps which floated downstream, probably for long distances and so were preserved in the deposits as alien waifs along with the far more abundant remains of entirely different types of the "local" lowland vegetation characteristic of the time.

PSILOPSIDA

Of these major groups of vascular plants, only the psilopsids did not develop arborescent members, perhaps because of an inherent inability to produce either a ring of secondary wood or a soundly functional root system capable of anchoring a heavy superstructure.

Aneurophyton

The exact affinity of this material has not yet been ascertained. Reproductively it was psilopsid in character and so usually has been classified with that group. However, it had considerable secondary growth and a good basal root system and hence became markedly arborescent; and it differed from typical psilopsids in other characters. Some years ago this plant received much publicity under the name of *Eospermatopteris,* the supposition being that it bore seeds. Investigations with newer techniques revealed that the so-called "seeds" found associated with extensive remains of it near Gilboa, in eastern New York State, actually were spore cases — and of another and apparently unrelated plant. With the discovery of the true nature of the trees making up the "Gilboa Forest," it became evident that *Eospermatopteris* was the same plant as the earlier described *Aneurophyton,* which name it now properly bears. As our knowledge of this ancient form increases, it may be found to have played a more important ancestral role than we have so far supposed. On the basis of modern concepts, *Aneurophyton* may yet be moved from its present position in the series to one perhaps between the pteropsids and cycadopsids.

Lycopsida

The four remaining genera of this major group tell us little of the grandeur of the Carboniferous lycopsid forests. The earliest known member of the group, *Baragwanathia,* so simple that sometimes it has been classed as a psilopsid, probably was not ancestral to the entire group; if anything, it gave rise to derived forms from which our modern lycopods have descended. The lepidodendrids, a motley group, had a series of components, only a few of which seem to have come directly from the geologically preceding protolepidodendrids; all these seemingly had an indirect origin out of a common phyletic plexus, perhaps during the Silurian.

The lepidodendrids, often with ample secondary growth and special systems of internal supporting tissues, were truly arborescent. Of interest, however, is the fact that both the aerial portions and root systems were dichotomously branched and similar in organization, except for such differences as we would expect between aerial and subaerial structures. Furthermore, the ultimate absorbing "rootlets" were coarse structures, seemingly little more than highly modified leaves. Hints of this old lepidodendrid system of organization still are retained in a few modern lycopsids such as *Isoetes.* The arborescent lycopsids were not engineered to meet the competition of the structurally and physiologically more efficient groups that were to come into prominence later during the Permian period.

Equisetopsida

The hyeniads, a curious group of smallish, almost shrublike plants, appear to be ancestral to the calamites and equisetums (the modern horsetails). The sphenophylls probably were a cognate group, at all times in their history rather weak and sprawling members of the swamp and, perhaps, moist upland communities. The arborescent calamites, abundant in the swamplands of the Carboniferous, had considerable secondary growth, yet in their organization and plant habit they were remarkably like our modern horsetails. These, in turn, may be little more than greatly reduced representatives of a collateral line developed from a basal member of the variable calamitean group no later than the Carboniferous and perhaps even in the Devonian.

Pteropsida

Although arborescent pteropsids were relatively abundant during the Carboniferous and since have been present in the several lines of this group, their arborescent nature always has been anomalous. The actual woody portion of their "trunks" never was large, the trees being primarily supported by a heavy mass of tightly intertwined roots which, originating at the bases of the leaves, grew downward, encasing the much weaker actual stem and thereby holding the plant erect. Nothing really new has happened in the basic architecture of the arborescent pteropsids since the Carboniferous. The development of numerous herbaceous and climbing pteropsids, as well as the derivation of forms suited to desert and aquatic habitats, may be merely noted.

Cycadopsida

The early history of our knowledge of this group is largely one of misunderstanding. For many years men have been astonished at the wealth of fernlike leaf impressions associated with Carboniferous strata. This easily led to the assumption, still maintained in some texts, that the Carboniferous was the "Age of Ferns." Coenopterid ferns were present, but for the most part they were both so uninteresting and so unfernlike as to be passed by. With the discovery that some of these fernlike leaves actually bore seed, a long controversy was started as to whether such forms were seed-bearing ferns, or seed plants with fernlike foliage.

Continued work on the group revealed that some never did bear seeds and also had the internal structure of true pteropsids. The bulk, however, appear to have been seed plants with fernlike leaves. More recent investigations, with improved techniques, have uncovered a large and complex series of these "seed-ferns," or pteridosperms and have also brought to light what, at last, appears to be excellent evidence that the majority were in no way related to the ferns but are close relatives of and in part ancestral to, the cycads. However, as we shall note later, a few of these so-called pteridosperms actually may be in the magnoliopsid line.

The cycadopsid pteridosperms were a varied lot. A number of them were arborescent, with abundant secondary growth in the roots, trunks and branches; some may have been of considerable stature. Others were shrubby, quite a few are known to have been high-climbing and anatomically complex woody lianas and there are hints that a number of herbaceous pteridosperms also were present. Long before they passed out of existence, the pteridosperms gave rise to both the cycads and cycadeoids.

Recent investigations of their anatomical structure indicate that various of the pteridosperms were capable of withstanding periods of drought. It may have been such forms which gave rise to the characteristically more xerophytic, arborescent cycads. Another derived group, the cycadeoids, differing from the cycads in various ways, seems to have been peculiarly adapted to semidesert conditions. These thick stemmed (but always lowgrowing) cycadeoids also evolved a "flower," but in its structure it was a cycadopsid flower and not one which could have given rise to the magnoliopsid flower. The few remaining species of the tropical, arborescent gnetums also appear to have been derived from some ancient group of pteridosperms.

Pinopsida: As here defined, the pinopsids encompass four collective subgroups: the extinct cordaites, the ginkgos, the "conifers" (some of which do not bear cones) and the taxads. The one living species of *Ginkgo* retains a primitive leaf structure and an even more primitive type of reproductive apparatus with swimming sperm; this latter feature also is characteristic of the cordaites and of all known "conifers" until after the Permian.

The taxads appear always to have stood somewhat apart from the other pinopsids, the wood structure of the group being the main clue to its basic affinity. To delve further into the intragroup relationships of the pinopsids in this place would open a controversy that is yet far from being resolved. But our uncertainty as to the exact relationships of the various pinopsid subgroups is a matter of insignificance in the face of the mystery of the origin of the group as a whole. In the upper reaches of Devonian strata, in deposits of estuarine and offshore nature, numerous trunk segments occur.

Microscopical examination of these, when well preserved, reveals that they constitute a series of pinopsid forms, some already with a highly evolved wood structure. All we can say is that these numerous trunk segments prove that the pinopsids are a very ancient group and that already in the Devonian the uplands of the world were clothed with pinopsid forests. From what we know of them, various of these ancient trees must have appeared much like modern araucarians.

Magnoliopsida: This group traditionally has been called the "angiosperms" (with covered seed), to distinguish it from the "gymnosperms" (with naked seed). A recent division of the so-called gymnosperms into *Cycadopsida* and *Pinopsida,* made necessary by their obviously different lines of descent, left the "angiosperms" standing unique without a name comparable to those of other groups. Therefore a widespread, well-known and relatively primitive genus, *Magnolia,* has been chosen to loan its name to this major group. The dicotyledonous hardwoods as well as the monocotyledonous palms constitute such important factors in our forest economy and are so familiar to all of us, that any discussion of their intragroup relationships in this place would be superfluous.

The often repeated concept that the magnoliopsids "came into being and underwent a great burst of evolution in the Cretaceous" is erroneous. Originally an upland group and therefore but rarely preserved in the earlier fossil record, the magnoliopsids found no opportunity until the middle portion of the Cretaceous to migrate in sufficiently large numbers into the lowlands to become commonly and widely preserved. When the group becomes at all commonly represented, we discover that already essentially all its modern arborescent families had been evolved and many of its modern genera also already had come into being, often with a series of species surprisingly similar to those of today. This indicates only that the magnoliopsids are not of recent origin, that they are no more "modern" than other groups of vascular plants. But the origin of the magnoliopsids is another matter.

Among the so-called pteridosperms may be found a series of forms which, even by the most conservative criteria, would be classed as aberrant for the group. In many ways these fit into our concepts of what, on purely theoretical grounds, we should expect to be forms ancestral to the magnoliopsids. But these aberrant pteridosperms were "gymnospermous" in their reproduction, whereas the magnoliopsids are typically "angiospermous" — and so the two have been kept separate by most workers.

But before we thrust these seemingly insignificant aberrant pteridosperms, with their seeds borne on modified leaves, back into the cycadopsids because of the gymnospermous nature of their reproductive apparatus, let us recall that there exists a series of magnoliopsid genera, admittedly primitive in the great magnolialean-annonalean complex of tropical, arborescent forms, which, even today, open their sometimes remarkably leaflike carpels at flowering time, thus exposing the naked ovules (the future seeds) in a typically "gymnospermous" form of pollination. We now suspect, also, that some of the wood fragments of the Devonian, currently assigned in a general way to the pteridosperms, on more careful examination may yet be found to be the wood of primitive gymnospermous members of the largely angiospermous magnoliopsids. A traditional passion for large and sometimes not too precisely defined terms in our so-called "scientific language" should not be a semantic block keeping us from an understanding of the true origins and developments of our great groups of vascular plants.

THE FIRST FORESTS

Direct fossil evidence of a vascular land flora appears for the first time in Silurian strata through a scattering of psilopsid forms and the primitive lycopsid *Baragwanathia.* Primitive as these forms appear to be in the light of what came after, they seem highly advanced when we consider the differences between them and any hypothetical, vascularized alga-like form from which they might have descended. There is considerable evidence that the first steps in the evolution of a group of organisms is a slow and halting process, accompanied by many excursions into biological blind alleys and beset with numerous failures. Therefore we can but conclude, on the basis of the known forms of the Silurian, that a vascular land flora of some sort already existed in the Ordovician and gave rise to that of the Silurian.

Unfortunately for us, the Silurian was a time of continental uplift and therefore one with few opportunities for the preservation of representative examples of what must have been an interesting succession of evolutionary experiments in the perfection of successful land-dwelling vascular forms. The Silurian culminated in a series of almost world-wide disturbances, accompanied by the building of great mountain ranges, which probably played much the same role that mountains always have in the development of vegetation types subsequently to become common and dominant in the lowlands.

The seemingly abrupt appearance, in the lowland deposits of the Devonian, of various forms of all major groups of vascular plants except the *Magnoliopsida* indicates that they must have been present and undergoing an active evolution on the uplands during Silurian time. There is a possibility that certain wood fragments of the Devonian, now assigned to the cycadopsid pteridosperms, actually may be of materials in the magnoliopsid line; if this proves to be so, the magnoliopsids would then have a recorded lineage just as ancient as that of any of the vascular groups which gave rise to arborescent forms. By Middle Devonian time the lands of the world were dominated by a series of forest types. Aneurophyton forests, mixed with gaunt-branched protolepidodendrids, were widespread in the lowlands; moist and swampy areas bore abundant psilopsids, hyeniads (the earliest known of the equisetopsids) and a variety of early coenopterid ferns.

The abundant segments of waterworn trunk and other wood fragments which began to appear in the deposits indicate the presence on the uplands of extensive pinopsid forests, probably with an understory of somewhat smaller pteridospermous trees and an undergrowth of lesser sorts.

TEMPERATE FORESTS

The transition from the tundra to temperate conditions in the northern hemisphere is characterized by the predominance of spruce-fir forest, called *taiga.* The southern tip of South America is too temperate to have a comparable forest type. On theoretical grounds the further transition southward from the taiga should witness a progressively greater incidence of broadleaf deciduous species; and in general this occurs.

In the North American Pacific region, however, the expected superhumid area, in conjunction with the mountains and generally higher elevations, has thrown the balance in favour of the retention of predominately coniferous forests, although many broadleaf genera and species are also found, especially in the lowlands. In the eastern part of North America, in Europe and in eastern Asia, the forests are primarily broadleaf deciduous, but coniferous elements of the north temperate forest also are abundant; this is the *Mixed Broadleaf-Coniferous Forest.*

The mixed broadleaf-coniferous forest of the southern hemisphere has much the same physiognomy as that of the northern, but with its own series of magnoliopsid and pinopsid genera. The superhumid region appears in southern Chile, greatly reduced in extent by the Andean chains; some of its characteristic species are closely related to forms living in New Zealand and other antipodean regions, or to forms known from Tertiary deposits of Antarctica.

In neither hemisphere is the transition from the temperate to the frostfree region marked by any sudden change in forest physiognomy, although the transition type is sometimes called *warm temperate forest.* This forest type often contains broadleaf evergreen species in genera which, in the more familiar temperate forest, are characteristically deciduous; broad-leafed evergreen pinopsids, unknown in northern regions, sometimes are found associated with this forest type in the southern hemisphere.

The occurrence of a fairly large number of relic genera in this transition zone may indicate that during the Early Tertiary this forest type was of much greater extent. This we should expect, for the generally warmer climate of the time would have produced a wider and even less clearly marked zone of transition between the temperate and frost-free regions.

ARID AND SEMIARID AREAS

The two central divisions of the drier areas of the world fall outside the scope of this work. The innermost, *Desert and Desert Scrub,* too dry to support anything more than a sparse vegetation, is encircled by short-grass *Steppe.* Surrounding the steppe, however, is a third segment which, vegetationally, is of interest here, for it contains arborescent material.

SUBHUMID TRANSITION AREAS

This outermost zone is transitional between grassland and forest. Near the western continental margins adjacent to or slightly within temperate regions, the rainfall occurs mostly in the winter months, the summer being hot and dry. Here a *Xeric Scrub and Woodland* is present, with trees along streams or in scattered, sparse colonies, usually accompanied by winter-green grasslands. This type of vegetation, best developed in the Mediterranean region (it is often called the "Mediterranean scrub" type) and in parts of California, also occurs in South America, southern Africa and Australia.

As this same transitional band swings toward the continental interiors, it becomes subject to violent storms and irregular precipitation. The trees are at first limited to stream valleys, forming "gallery forests". With slightly more precipitation, groves of trees and thickets of shrubs begin to appear away from streams.

In the mid-continent of North America, the groves were once largely dominated by oaks, often bur oak *(Quercus macrocarpa)* and offered a welcome relief to those crossing the grasslands during the period of early settlement; stag-headed relics of some of these oak groves still stand near old homesteads, but most of them have been removed because of the danger of storm-thrown branches. In North America the shrub thickets often were dominated by thorny species of plum *(Prunus)* and hawthorn *(Crataegus).* The intervening grasslands contained relatively tall grasses and have been called *Prairie.* It is here suggested that the term "prairie" might perhaps be expanded to take in the entire vegetation complex of the region.

In its undisturbed phase the prairie is a transition grassland containing gallery forest and occasional groves of widely spaced trees, usually associated with more or less thorny shrubs. This is characteristic of the Asiatic, Eurasian and South American prairies as of those of midcontinent North America. Today there is little opportunity anywhere in the world to see typical prairie.

Climatically suited to the raising of excellent grain crops, for the grains are grasses, the natural prairie areas of the world are almost entirely under intensive cultivation, so great is the world's need for food. The gallery forest and scattered groves have largely fallen because of acute needs for fuel and local building material, although there is evidence that many of the groves and some of the thinner gallery forests were eliminated by prairie fires set by indigenous nomadic hunters too early to have been seen by travelers in modern times. The "treeless prairie," in large part, may be the result of clearing and fire.

Extending into the frost-free areas, the "tropical" margin of this subhumid region becomes *Savanna* at about latitude 30°, perhaps more by definition than by any immediately visible change in vegetation. Here the precipitation strongly tends to be concentrated in the high-sun months, although there may be a second period of low-sun rainfall. The savanna is characteristically a tropical grassland, often disturbed by fire, with gallery forest along the streams and with scattered groves or occasional individual trees. Palms are not infrequent and the shrubby growths tend to be thorny. Typical *Thorn Scrub* and *Thorn Forest* usually appear adjacent to either savanna, or light tropical forest and often develop out of these types after disturbance.

TROPICAL FORESTS

Primary tropical forest is not easy to find, although much has been so labeled by certain writers. In his excellent treatise on the world's tropical rain forests, Richards adds a most revealing footnote. In discussing what, in the original text, he characterizes as "Primary Mixed Forest" in Nigeria, he remarks that further work in the area "makes it seem likely that this forest has suffered disturbance in the past and is old secondary rather than truly primary." He adds: "The same is probably true for nearly all so-called virgin or primary forest in Nigeria and perhaps the whole of West Africa." Undisturbed primary forest is usually thought to be of considerable extent in the American tropics; but it is by no means so widespread as is generally supposed.

THE TROPICAL RAIN

Forest is perhaps our least understood vegetation type. Many forest areas in the tropics classed as "rain forest" actually are *Light Tropical Forest.* A further complication is the presence of large forest tracts on mountain slopes in the tropics which, because of the terrain and wind pattern, are well watered. These are *Tropical Montane Rain Forests;* on the middle slopes where the precipitation is usually highest, they have a composition considerably different from that of the rain forests of the lowlands. Where they are adjacent, both lowland rain forest and light forest pass into the montane rain-forest type.

However, if we take a rather narrow zone across the equator, including those areas in which rain forest would be expected to develop, we discover that it is advisable to particularize our information. It is immediately obvious that there is only a narrow strip along the equator in which rain can be expected to fall at almost all times of the year. Here seasonal maxima may occur, but only exceptionally does any long period pass without rain. This relatively narrow equatorial band is the locale of the true rain forest, although local conditions in some places produce equivalent conditions that enlarge the occurrence of this forest type. Furthermore, some of the areas commonly mapped as rain forest actually contain considerable tracts of light forest.

Wet and dry seasons begin to be evident at only a few degrees of latitude north and south of the equator. The total annual rainfall may be very great, but, on the whole, true rain forest does not exist where the dry seasons last longer than a few weeks. However, in such areas along the main river valleys and even along the smaller streams, the residual soil moisture is such that what might reasonably still be classed as rain forest does occur.

This readily explains why observers traveling only along streams have in the past often expanded the supposed general distribution of this type of forest far beyond its actual limits. Glimpses of forest were seen in the distance and it was supposed that it was of the same general composition as that near at hand. Actually this riparian "rain forest," intercalated with light forest, is comparable to the fingers of gallery forest that follow the streams in savanna regions.

Light forest is found in those regions near the equator where there are two yearly maxima of rain, with one of the dry seasons of about three months in length. With a shortening of the longer dry season, this forest type approaches the rain forest in character.

However, on well-drained to overdrained terrain, the amount of moisture retained-the "effective precipitation" — may be low enough to produce characteristic light forest relatively close to the equator (and, in fact, in some places actually athwart it). Where the rainfall pattern is such that there is only one rainy period and a correspondingly long dry period, light forest persists until the conditions are reached where natural savanna (or in some instances thorn forest and scrub) occurs.

The light forest is obviously not a single forest "type." It may be classified on a local basis, but in broad pattern it varies from a "wet" phase to a "dry" phase through myriad nuances of vegetational association. In discussing both the rain forest and what actually is the "wetter" phase of light forest, Richards indicates something of its nature. There are exceedingly complex ("mixed") types, with no species really dominant in the stands and there are also extensive forests with single dominant species.

He likens the "mixed" tropical forest to the mixed mesophytic forest of eastern North America, which develops only in areas with abundantly varied micro-environments. I suggest here that the "mixed" forest of the tropics may also be a type that develops primarily in areas with a series of varied micro-environments, as yet essentially unstudied and therefore not understood. The forests near major streams, where they are subject to periodic, or occasional, inundation — areas often many miles in width — are, of course, "disturbed forests" and therefore likely to be complex, with a characteristic mixed community and a considerable amount of "jungle" undergrowth.

In the light forest, unless disturbed in some manner, the canopy is usually sufficiently close to prevent herbaceous and shrubby forms from receiving light enough to form more than a hint of jungle-like undergrowth. On the whole, agriculture in the tropics is not successful in those areas characterized by typical rain forest, nor is life easy there. Hence, in his constant search for habitable places, man long ago entered the light forest, cleared the land and set about his routine chores of cultivating it. Great areas in India, once covered with "monsoon forest" (a typical form of light forest produced by the strong periodicity of rainfall), are now given over to intensive agriculture. The more primitive pastoral agriculture of central Africa has also eliminated large stretches of light forest, but here savanna has resulted, so that it is today most difficult to distinguish between natural savanna and that which has been artificially produced by fire and overgrazing.

Great as are the forest resources of the tropics, we have as yet made far too little intelligent use of them. At present, the world's primary demand is for the coniferous softwoods. Numerous especially valuable timbers are today being cut from tropical forests, but the great majority of the species are ignored. Remote from the daily experience of most of us, the spoilage of the remaining areas of tropical rain forest and light forest through improper practices is not generally comprehended.

Under our present economy many of the species of these forests are considered "worthless" and their normal habitats are sometimes almost wantonly destroyed. These "worthless" forms constitute at least a vast reservoir of raw cellulose; their other uses may be legion, should we study their characteristics in greater detail. Here is a great natural resource rapidly being wasted away through ignorance.

Variability and Kind

The origin of forest types, those times in the history of the world when the land surfaces were elevated, when mountain building was at a maximum and when climatic disturbances were widespread marked important periods in the development of new vegetation types. In view of the changes that land surfaces and climates constantly undergo, a too rigid genetic system is a distinct disadvantage to any group of organisms. With genetic plasticity — variability -there is a greater chance of survival during any disorganization of habitat. Variability, indeed, is the basis of survival.

From the standpoint of plant systematics the species is usually considered to be the working unit. Furthermore, it is traditionally supposed to exhibit but little variation. This concept, however, is largely based on superficial morphological grounds.

As we begin to study species in greater detail, examining thousands of individuals instead of only a few in each species, it becomes evident that even this so-called morphological constancy of the species has little basis in fact; we have not been scrutinizing our materials with sufficient precision. We also are learning that, with this morphological variability, there is an equally large amount of physiological variability, inherent in the genetics of the group; and, in the end, this inherited physiological variability is far more important than are differences in morphology where factors concerned with survival are involved.

Foresters have long known that, in reforestation or afforestation, it is best to use seed from sources not too far removed from the site of the new plantings. In western North America Douglas fir (*Pseudotsuga taxifolia*) grows from about latitude 55° in British Columbia to about latitude 170 in Mexico, from sea level in the Pacific Northwest to elevations of over 10,000 feet in the Rocky Mountains and from humid to subhumid habitats.

Attempts have been made to set up a series of "segregate species" from within the known morphological variability of this wide-ranging tree, but from the point of practical botanical taxonomy only the variety *glauca* is now recognized apart from the "typical" form. However, it has been found best not to move materials much more than a degree of latitude, or a climatically comparable distance in elevation, if the greatest survival and maximum growth of the planting is desired, though practical considerations usually make this impossible to achieve.

As a species defined by taxonomists, Douglas fir is widespread, but each of its local biotypes has a genetic norm that fits it to a particular set of ecological conditions. Even here we can discover considerable variability, related to the precise conditions of the local micro-environments. There also is a growing body of data that day length at certain periods of the year — which, of course, is correlated with latitude — plays a very important part in the effective competition and survival of these minor biotypes.

These and similar examples might be greatly expanded. The white pine (*Pinus strobus*) is found from eastern Canada through the Appalachian Mountains and also in the highlands of southern Mexico. When brought together at about latitude 40° N and if protected by appropriate handling under glass, the material from Mexico grows considerably faster than does that from the central Appalachians, as in turn this grows faster than do the Canadian plants. When planted in the open at this same latitude, the more nearly local material still grows faster than does the Canadian material, but the Mexican plants die with the first hard freeze. Data of the same sort are being accumulated for hardwoods that have wide distributions and also apply to adaptability to variant soils as well as to differences in climate-in brief, to the host of factors that influence the growth and persistence of plants.

The source of this genetic variability within species and the mechanics of its operation within species populations, which often enables the sum of their biotypes to have considerable geographical ranges and ecological tolerances, is a subject too extensive to be developed here. Nevertheless, what we can ascertain by relatively precise experimental methods with regard to what is happening among the species around us today, is indicative of what has been going on through the long span of time during which forests have covered the land.

Through such studies we come to an understanding of how it was that, in the past, groups of organisms could develop biotypic variations capable of venturing into new habitats and new climatic regions.

Thus we glean insight into the importance of the uplands and mountain ranges of the past. It was on the uplands and cooler mountain slopes of the frost-free regions that the ancestors of our modern forest trees evolved. On these same slopes, with their multitude of different micro-environments, variant forms better preadapted to conditions in temperate regions came into being, were further elaborated, could persist for long periods of time and so be available to spread poleward during those periods when the great cyclic changes in the topography of the continental masses and their climates afforded an opportunity.

Likewise, this same inherent tendency toward variation, characteristic of all life, gave rise to forms still better fitted to cooler or even to boreal climates. With a swing of the climatic pendulum in the other direction, the ancestral biotypes would necessarily have had to retreat, or, if unable to do so for some reason, become extinct, leaving behind the newly evolved forms. These derived forms are the backbone of the vegetation of our present temperate and cool-temperate forests. With further evolution, coupled with genetic isolation, these biotypic segregates could — and have — become the "boreal" and "extreme austral" species of the taxonomist.

There is a tendency of some people to think of the tropical lowlands, because of their generally equable conditions, as having had a persistent and stable series of forest communities over long periods of time.

This would seem to be a valid assumption on theoretical climatological grounds and also because considerable segments of the equatorial lowlands have been in a reasonably stable geographical position, at least since the Permian. However, there is evidence that the present forest types of the lowlands have been derived in large part from more upland types, which migrated into the lowlands during the Middle and Late Tertiary. As the writer has pointed out elsewhere, the forests of the central Amazonian basin appear to be primarily composed of elements derived from the surrounding uplands, perhaps at a time no earlier than the Late Pliocene or Early Pleistocene.

Obviously, based as it is on the consideration of a single area, this generalization needs considerable study before it can be taken as being more widely applicable. On the whole, however, the forests of tropical lowlands do not contain what, according to generally accepted botanical criteria, are considered to be the more primitive living species of genera that have fairly broad altitudinal distributions. These are usually found on the piedmont and lower mountain slopes. Proper utilization of our forest resources demands that trees be cut down. What comes back to take their place-or what is brought in to replace a forest area that has been despoiled by improper cutting practices — is not a thing to be decided in a casual manner.

Even with selective thinning, the micro-environments have been changed and with clear-cutting a totally new set of factors is introduced. Some few forest types regenerate satisfactorily; others are replaced — or might better be replaced — with different sets of materials. Whether the replacement is to be with the same biotypes, with somewhat different biotypes of the original species, or with entirely different species, depends on a considerable array of factors.

These factors are within the province of scientific silviculture, but they cannot be ascertained without a great deal of painstaking research.Biotypic variation is the fundamental mechanism leading to the evolution of the manifold kinds of organisms with which this world abounds and an understanding of it is indispensable in any consideration of the proper utilization and regeneration of our forests, our most important replaceable natural resource.

FORESTS OF FROST-FREE REGIONS

Despite the larger land areas in the northern hemisphere, more than half of the world's remaining forests are to be found in frost-free regions. In the higher mountains of the tropics, especially where contiguous to the temperate regions, the vegetation is likely to contain genera often thought to be "characteristic" of the temperate forest, commingled with elements of more "tropical" groups. Here the rapid changes in elevation, coupled with the great number of micro-environments on rugged slopes, lead to a complex type of vegetation.

At middle elevations, the forests merge into those characteristic of the adjacent tropical lowlands. In a work of this compass it would be useless to attempt any description of these frost-free montane forests. Where the precipitation is abundant-as it often is, especially on the windward sides — a *Montane Rain-Forest Complex* develops, with the dominant species often succeeding each other in relatively narrow altitudinal belts; it is "complex" because each species is present but progressively less conspicuous in a succession of adjacent zones.

DISTURBED TROPICAL AREAS AND "JUNGLE."

There is a growing body of evidence that many areas of present thorn forest and related thorn scrub have developed out of both savanna and light tropical forest following disturbance, especially overgrazing. The great development of tropical savanna in many areas, however, has been accepted as a natural phenomenon by many plant geographers. Natural savanna would be expected to occur in the climatic tension zone between tropical steppe and light forest. Nevertheless, on critical examination, it becomes evident that there is a proportional imbalance between savanna and light forest in various regions.

While this is no more true of parts of Africa than of other tropical areas, it is most evident there, with far more savanna and much less light forest than we should expect on theoretical grounds. The often sharp line between relatively heavy forest and savanna, with no apparent differences in soil or precipitation, indicates that the sudden change from one vegetational type to the other is unnatural. There now is a strong tendency among recent students to conclude that wide expanses of savanna, as now defined, have been caused by fire. If so, the question is whether this is a recent happening, or whether — in some areas — it may be a legacy of the antique past. Again we may turn to Africa for evidence, evidence that has been available for a long time, but has been largely dismissed as distorted folk lore.

The *Periplus* of Hanno, the Carthaginian, the Greek translation of which still survives, is an account of a remarkable voyage of exploration along the west coast of Africa about the year 520 B.C. Hanno relates that some time after passing the mouth of the Gambia River he stopped on a coastal island, where the night was made hideous by the sound of drums, gongs and some sort of wind instrument and terrifying by great fires.

Leaving the place in haste, Hanno sailed on, keeping Africa on his left. Day after day columns of smoke rose from the land and by night the whole country seemed to be in flames. Whether the explorers got no farther than Nigeria, or ventured to the Cameroons, as some maintain, is of little moment here. The main point for this discussion is that Hanno recorded having seen great fires inland in Africa for days on end as he sailed along its coast.

Although Hanno failed to comprehend their nature, these fires must have been due to the annual "burning of the bush" to beat back the forest and provide land for agriculture and grazing. Thus we have evidence that for at least two and a half millennia — and how much longer we shall never know — great forest tracts in Africa have probably been regularly despoiled by fires purposely set by man, much as they are today. The result has been a great expansion of savanna at the expense of the seasonally combustible light forest; and the same thing has unquestionably happened over large areas in the Americas, in India and in southeastern Asia.

Only within recent years have students working in tropical areas been able to divorce their thinking from ideas gleaned from previous study of the exceedingly overpopulated temperate regions and of the impoverished floras with which they are already familiar. Simplified concepts of forest ecology based on the almost diagrammatic floristic pattern of northeastern North America, or especially of northwestern Europe with its relatively few forest species, do not readily apply to the tropics, nor are the farms and pasture lands of the tropics laid out in neatly fenced plots. The number of species in a given area usually is far greater in tropical than in temperate regions; and agriculture in many parts of the tropics is fundamentally migratory, for several very definite reasons. In the drier portions, catch crops planted during, or immediately following, the rains are the rule. Here clearing is primarily by fire and the fires, set in the periods between rains, often get out of hand, spread over great areas and enter the forest margin for some distance. This produces the often observed sharp line of demarcation between forest and savanna. It does not take a particularly dense human population to bring about these effects and man has been living in the equatorial regions of both hemispheres for much longer than is sometimes supposed.

In the lowland portions of the frost-free zone that have abundant precipitation-in rain forest and light forest-there is and for a very long time has been, a considerable population, especially near the watercourses, the "jungle roads." Here patches are cleared for the indigenous crops (and more recently for those that have been introduced) and the land is cultivated. Again, however, one familiar only with agriculture in temperate regions is prone to misinterpret the situation; the native farmer is not being indolent when he clears a plot of ground, cultivates it for a few years and then abandons it to weeds. It is the only course open to him.

The generally high soil temperatures do not lead to humus formation and soon the abundant rainfall so thoroughly leaches the soil of its mineral nutrients that it is necessary to abandon the plot after only a few crops. A new area is cleared, the rubbish burned where it lies (an effective method of scattering the mineral-containing ash) and a new crop planted. Unfortunately, modern plantation agriculture in the tropics far too often follows this same pattern. Under this system "fallowland" is a former cultivated area, perhaps only temporarily abandoned as no longer profitable because of the dearth of nutrient minerals in the soil and the high cost of fertilizers.

After abandonment, these cleared and briefly cultivated areas return to vegetation natural to the area, but not to the original forest type. Such tropical "weed forest" is the most rampantly growing sort of vegetation known. But remarkable as is this secondary forest, with its dense undergrowth and tangle of lianas, it is characteristic only of the disturbed areas of the region.

Descriptions of these "impenetrable jungles," usually written in awed and sometimes lurid phrases by early travelers, who necessarily had to stay close to the rivers, have become the classic texts from which we have taken our concepts of tropical vegetation for much too long a time. Breaking out of this river-margin "jungle" and complex secondary growth that follows cultivation, the traveller may walk, as the writer has, day after day in the almost trackless forest of the South American equatorial region, only rarely having to swing his machete-except, if he be a botanist, to collect some choice specimen. The undisturbed tropical forest may be complex and can contain a great abundance of species, but it is far different from the "jungle" of much of our literature.

CLIMATE CHANGE IMPACTS ON FOREST HEALTH

The world's climate is changing. Increased temperatures and levels of atmospheric carbon dioxide as well as changes in precipitation and in the frequency and severity of extreme climatic events are just some of the changes occurring. These changes are having notable impacts on the world's forests and the forest sector through longer growing seasons, expansion of insect species ranges, and increased frequency of forest fires. Climate change can affect forest pests and the damage they cause by: directly impacting their development, survival, reproduction and spread; altering host defences and susceptibility; and indirectly impacting the relationships between pests, their environment and other species such as natural enemies, competitors and mutualists. A deeper understanding of the complex relationships between a changing climate, forests and forest pests is vital to enable those in forest health protection and management to expect and prepare for changes in pest behaviours, outbreaks and invasions. This chapter investigates these relationships and the implications for forest health protection and management.

Specific examples where insect and pathogen lifecycles or habits have been altered by local, national or regional climatic changes are discussed. For example, the mountain pine beetle (*Dendroctonus ponderosae*) has shown decreases in generation time and winter mortality resulting in exponential population growth and major range extension in western North America. Warmer temperatures have resulted in range expansions of pests such as the pine processionary caterpillar (*Thaumetopoea pityocampa*) and the oak processionary caterpillar (*T. processionea*) in Europe and the southern pine beetle (*D. frontalis*) and red band needle blight (*Mycosphaerella pini*) in the US. Similar patterns are expected for other pathogens such as *Armillaria mellea,* and *Phytophthora cinnamomi* in Europe.

In the Democratic Peoples Republic of Korea, decreased winter temperature and lighter snow cover have contributed to reduced mortality and early emergence of overwintering *Dendrolimus spectabilis* (pine moth) with resulting loss of native *Pinus densiflora.*

The purpose of this report is to compile and summarize the present knowledge on the impacts of climate change on forests and forest pests. Since limited research has been dedicated to determining such impacts on forest pests in particular, information on non-forest pests is provided throughout the report to enable a better understanding of the potential impacts of climate change on forest health.

CURRENT KNOWLEDGE AND FUTURE EXPECTATIONS

The world's climate is changing. While there are natural climate variations, the climate change we are most concerned about is in regards to the modifications to the greenhouse effect that human activities have caused. The Intergovernmental Panel on Climate Change (IPCC), in its Fourth Assessment Report, concluded with more certainty that global climate change is unequivocal and it is widely believed to result primarily from the effects of emissions of carbon dioxide (CO_2) and other greenhouse gases (GHGs) such as methane (CH_4) and nitrous oxide (N_2O), from human activities.

This conclusion was based on a number of observations about the Earth's climate including the following.

- Global surface temperature has increased by an estimated 0.74 degrees Celsius (°C) over the past century. Over a 50 year period from 1956 to 2005 the warming trend was nearly twice that for the 100 years from 1906 to 2005. Eleven of the 12 years from 1995 to 2006 rank among the 12 hottest years on record (since 1850, when sufficient worldwide temperature measurements began). Increases are widespread globally and are greater at higher northern latitudes. Over the last 50 years, cold days and nights as well as frosts have become less frequent over most land areas, while hot days and nights and heat waves have become more frequent.
- Consistent with warming, the extent of snow and ice has decreased. Mountain glaciers and snow cover have also declined on average worldwide. The maximum area of seasonally frozen ground in the Northern Hemisphere has decreased by about 7 percent since 1900, with decreases in the spring of up to 15 percent. Satellite data since 1978 show that the extent of Arctic sea ice during the summer has shrunk by more than 20 percent.
- Since 1961, the world's oceans have been absorbing more than 80 percent of the heat added to the climate, causing ocean water to expand thereby contributing to rising sea levels. This expansion was the largest contributor to sea level rise between 1993 and 2003. Melting glaciers and losses from the Greenland and Antarctic ice sheets have also contributed to recent sea level rise. The incidence of extreme high sea level has increased at a number of sites globally since 1975.
- From 1900 to 2005, significant increases in precipitation have been observed in eastern parts of North and South America, northern Europe and northern and central Asia. The frequency of heavy precipitation events has also increased over most areas. In contrast, precipitation has declined in the Sahel, the Mediterranean, southern Africa and parts of southern Asia. Droughts have become longer and more intense worldwide and have affected larger areas since the 1970s, particularly in the tropics and subtropics.
- Evidence of an increase in intense tropical cyclone activity in the North Atlantic has been observed since about 1970 and there are suggestions of similar increased activity in some other regions.

Predictions for the Future

The IPCC has made a number of predictions for future climate change. A warming of about 0.2°C per decade is projected for the next two decades; temperature projections beyond this period depend on specific emissions scenarios. Even if the concentrations of all GHGs and aerosols were kept constant at year 2000 levels, a further warming of about 0.1°C per decade would be expected.

The full range of projected temperature increase, based on six emission scenarios, is 1.1-6.4 °C by the end of the century. The best estimate range of projected temperature increase, which extends from the midpoint of the lowest emission scenario to the midpoint of the highest, is 1.8-4.0 °C by the end of the century. It is very likely that continued GHG emissions at or above current rates would cause further warming and induce many changes in the global climate system during the 21st century that would be larger than those observed during the 20th century.

Geographically the predicted warming trends for the 21st century are believed to be similar to those observed over the past several decades whereby temperature increases are expected to be greatest over land and at most high northern latitudes, and least over the Southern Ocean (near Antarctica) and the northern North Atlantic.

It is very likely that extreme heat, heat waves and heavy precipitation events will become more frequent. Increases in high latitude precipitation are very likely, while decreases are likely in most subtropical regions such as Egypt. It is also likely that future tropical cyclones (typhoons and hurricanes) will become more intense, with higher peak wind speeds and heavier precipitation associated with warmer tropical seas. Snow cover area is projected to contract, widespread increases in thaw depth are projected over most permafrost regions and sea ice is projected to shrink in both the Arctic and Antarctic. In some projections, Arctic late-summer sea ice disappears almost entirely by the latter part of the 21st century.

Climate change is impacting the world's ecosystems and it is expected that the magnitude of these impacts will increase along with temperatures over this century. Many species and ecosystems may not be able to adapt as the effects of global warming and associated disturbances, such as floods, drought, wildfire and insect outbreaks, are compounded by other stresses such as land use change, overexploitation of resources, pollution and fragmentation of natural systems. If the global average temperature increases more than 1.5-2.5 °C, it is believed likely that approximately 20-30 percent of plant and animal species assessed so far will be at an increased risk of extinction. Major changes in ecosystem structure and function, species' ecological interactions and species' geographical ranges, with predominantly negative consequences for biodiversity and ecosystem goods and services are also projected.

IMPACTS ON FORESTS AND THE FOREST SECTOR

Climate change, in particular increased temperatures and levels of atmospheric carbon dioxide as well as changes in precipitation and in the frequency and severity of extreme climatic events, is having notable impacts on the world's forests and the forest sector.

Productivity and Health

Forest productivity and species diversity typically increase with increasing temperature, precipitation and nutrient availability, although species may differ in terms of their tolerance. As a key factor that regulates many terrestrial biogeochemical processes, such as soil respiration, litter decomposition, nitrogen mineralization and nitrification, denitrification, methane emission, fine root dynamics, plant productivity and nutrient uptake, temperature changes are likely to drastically alter forests and ecosystem dynamics in many ways. The impacts of elevated temperatures on trees and plants will vary throughout the year since warming may relieve plant stress during colder periods but increase it during hotter periods.

Moisture availability in forests will be strongly influenced by changes in both temperature and precipitation. Warmer temperatures lead to increased water losses from evaporation and evapotranspiration and can also result in reduced water use efficiency of plants. Longer, warmer growing seasons can intensify these effects resulting in severe moisture stress and drought. Such conditions can lead to reductions in the growth and health of trees although the severity of the impacts depends on the forest characteristics, age-class structure and soil depth and type. Young small plants such as seedlings and saplings are particularly susceptible whereas large trees with a more developed rooting system and greater stores of nutrients and carbohydrates tend to be less sensitive to drought, though they are affected by more severe conditions. Shallow-rooted trees and plants as well as species growing in shallow soils are more susceptible to water deficits. Deep-rooted trees can absorb water from greater depths and therefore are not as prone to water stress. Moisture stress and drought can also impact forest health by enhancing susceptibility to disturbances such as insect pests and pathogens and forest fires.

Higher atmospheric CO_2 levels result in increased growth rates and water use efficiency of plants and trees, so long as other factors such as water and nutrients (*e.g.* nitrogen, phosphorus, sulphur, some micronutrients) do not become limiting. It has been suggested however that this positive effect declines with increasing concentrations (Stone, Bhatti and Lal, 2006). However current free air CO_2 enrichment facilities have observed multi-year growth increases of 23 percent with a 175 ppm enrichment above a 375 ppm ambient CO_2 concentration. Elevated carbon dioxide levels can also result in changed plant structure such as increased leaf area and thickness, greater numbers of leaves, higher total leaf area per plant, and larger diameter stems and branches. It is important to note that plant responses to CO_2 enrichment may differ between species and local environmental conditions which are likely to result in substantial changes in the species composition and dynamics of terrestrial ecosystems. Concurrent increases in concentrations of ground-level ozone (O_3) may lower tree productivity and enhance susceptibility to pathogens , while N_2O may enhance growth in nitrogen-limited ecosystems such as boreal forests.

Distribution

Consistent responses of species and communities to climate change or 'fingerprints' are typically associated with changes in their distribution, particularly at their latitudinal or altitudinal extremes. The distribution of forest plants and trees is expected to shift northwards or to higher altitudes in response to climate warming.

In a recent study, Lenoir *et al.* (2008) compared the altitudinal distribution of 171 forest plant species between two periods 1905-1985 and 1986-2005 in Western Europe and concluded that climate warming has resulted in a significant upward shift in species optimum elevation (altitude of maximum probability of presence) averaging 29 m per decade. This study showed that climate change affects not just species' ranges at their distributional margins but at the spatial core of the distributional range of plant species. The quickest species noted to relocate were those with shorter life spans and faster reproduction cycles such as herbs, ferns and mosses; larger long-lived trees and shrubs did not show a significant shift and are thus under greater threat from the impacts of climate change because they can't adapt to local conditions quickly enough and relocate. Such distributional changes will no doubt result in forest ecosystems that are very different from what we see today. A similar study from 26 mountains in Switzerland reported that alpine flora have expanded toward the summits since the 1940s. Upward movements of tree lines have also been observed in Siberia, the Canadian Rocky Mountains, and New Zealand and northward shifts have been noted in Sweden and Eastern Canada.

The timing of such shifts however will not be solely determined by temperature but will depend on a number of factors such as the rate at which seeds can disperse into new regions that are climatically suitable (*i.e.* with proper moisture conditions, soil characteristics and nutrient availability), possible human interventions to promote movement of species, and changes to disturbance regimes.

Forest Sector

All of these impacts on trees and forests will inevitably have widespread impacts on the forest sector. Changes in the structure and functioning of natural ecosystems and planted forests (due to temperature changes and rainfall regimes) and extreme events and disasters (hurricanes, droughts, fires and pests) will have negative impacts on the productive function of forest ecosystems which in turn will affect local economies. Production patterns and trade in forestry commodities will be altered as species are grown more competitively in higher latitudes and altitudes.

Conversely, markets may be saturated due to increased mortality of trees following pest infestations as has been experienced with the mountain pine beetle in Canada. Decreased forest ecosystem services, especially water cycle regulation, soil protection and conservation of biological diversity, as a result of climate change may imply increased social and environmental vulnerability.

While climate change is likely to increase timber production and lower market prices in general, the increases in production will certainly not be evenly distributed throughout the world; some areas will experience better conditions than others. For example, forests with low productivity due to drought will likely face further decreases in productivity, while areas where temperature limits productivity may benefit from rising temperatures.

Disturbance

Forests are subjected to a variety of disturbances that are themselves strongly influenced by climate. Disturbances such as fire, drought, landslides, species invasions, insect and disease outbreaks, and storms such as hurricanes, windstorms and ice storms influence the composition, structure and function of forests. Climate change is expected to impact the susceptibility of forests to disturbances and also affect the frequency, intensity, duration, and timing of such disturbances. For example, increased fuel loads, longer fire seasons and the occurrence of more extreme fire weather conditions as a consequence of a changing climate are expected to result in increased forest fire activity. A changing climate will also alter the disturbance dynamics of native forest insect pests and pathogens, as well as facilitating the establishment and spread of non indigenous species.

Such changes in disturbance dynamics, in addition to the direct impacts of climate change on trees and forest ecosystems, can have devastating impacts particularly because of the complex relationships between climate, disturbance agents and forests. Either of these disturbances can result in forest susceptibility to other disturbances.

For example, pine forests in Central America became infested with bark beetles, primarily *Dendroctonus frontalis* in association with other *Dendroctonus* and *Ips* species, after suffering damage from Hurricane Mitch in 1998. The beetle outbreaks caused further extensive tree mortality thereby increasing fuel loads in the region's forests which severely increased the risk of wildfires. Making predictions on the future impacts of a changing climate on forest disturbances is made more difficult by these interactions.

IMPACTS ON INSECTS AND DISEASES

Changes in the patterns of disturbance by forest pests (insects, pathogens and other pests) are expected under a changing climate as a result of warmer temperatures, changes in precipitation, increased drought frequency and higher carbon dioxide concentrations. These changes will play a major role in shaping the world's forests and forest sector.

Climate change can affect forest pests and the damage they cause by: directly impacting their development, survival, reproduction, distribution and spread; altering host physiology and defences; and impacting the relationships between pests, their environment and other species such as natural enemies, competitors and mutualists.

Direct Impacts

Climate, temperature and precipitation in particular, have a very strong influence on the development, reproduction and survival of insect pests and pathogens and as a result it is highly likely that these organisms will be affected by any changes in climate. With their short generation times, high mobility and high reproductive rates it is also likely that they will respond more quickly to climate change than long-lived organ-isms, such as higher plants and mammals and thereby may be the first predictors of climate change.

Physiology

Climate influence on insects can be direct, as a mortality factor, or indirect, by influencing the rate of growth and development. Some information on the impacts of increased CO_2, and O_3, is becoming available but only for specific environments and only very partial information is available on changing UVB levels and altered precipitation regimes. For these reasons this report will focus on the impacts of temperature. Temperature is considered to be the more important factor of climate change influencing the physiology of insect pests. Precipitation however can be a very important factor in the epidemiology of many pathogens, such as *Mycosphaerella pini*, that depend on moisture for dispersal.

The magnitude of the impacts of temperature on forest pests will differ among species depending on their environment, life history, and ability to adapt. Flexible species that are polyphagous, occupy different habitat types across a range of latitudes and altitudes, and show high phenotypic and genotypic plasticity are less likely to be adversely affected by climate change than specialist species occupying narrow niches in extreme environments.

Increases in summer temperature will generally accelerate the rate of development in insects and increase their reproductive capacity while warmer winter temperatures may increase overwinter survival. Perhaps the best example of such impacts is the mountain pine beetle (*Dendroctonus ponderosae*) which has been at epidemic proportions in western Canada for several years.

Successive years of mild winters have decreased the mortality of overwintering stages and generation time allowing for massive destruction of pines in the region, particularly lodgepole pine (*Pinus contorta*). Decreased snow depth associated with warmer winter temperatures may also decrease the winter survival of many forest insects that overwinter in the forest litter where they are protected by snow cover from potentially lethal low temperatures.

The impact of a change in temperature will vary depending on the climatic zone. In temperate regions, increasing temperatures are expected to decrease winter survival while in more northern regions, higher temperatures will extend the summer season thereby increasing growth and reproduction. Given the more severe environmental control, and greater predicted increases in temperature in boreal and polar regions, the impacts of temperature are expected to be greater on species from those regions than on species in temperate or tropical zones. However, Deutsch *et al.* (2008) suggest that, in the absence of ameliorating factors such as migration and adaptation, the greatest extinction risks from global warming may be in the tropics. Warming in the tropics, though proportionately smaller in magnitude, could have the most deleterious impacts because tropical insects have very narrow ranges of climatic suitability compared to higher latitude species, and are already living very close to their optimal temperature.

Some important forest insect pests have critical associations with symbiotic fungi but limited information is available on how temperature changes may affect these symbionts and thus indirectly affect host population dynamics. In some cases insect hosts and their symbionts may be similarly affected by climatic change while in other cases, hosts and symbionts may be affected asymmetrically, effectively decoupling the symbiosis.

Distribution

Climate plays a major role in defining the distribution limits of a species. With changes in climate, these limits are shifting as species expand into higher latitudes and altitudes and disappear from areas that have become climatically unsuitable. Such shifts are occurring in species whose distributions are limited by temperature such as many temperate and northern species. It is now clear that poleward and upward shifts of species ranges have occurred across many taxonomic groups and in a large diversity of geographical locations during the 20th century. Parmesan and Yohe (2003) reported that more than 1 700 Northern Hemisphere species have exhibited significant range shifts averaging 6.1 km per decade towards the poles (or 6.1 m per decade upward).

The range expansions of many lepidopterans have been particularly well documented. Parmesan *et al.* (1999) reported a poleward shift of 35-240 km for 22 out of 35 non-migratory European butterfly species during the last century. Wilson *et al.* (2005) noted that the lower elevational limits of 16 butterfly species in central Spain had risen on average by approximately 212 m in 30 years, a rise attributed to an observed 1.3 °C rise in mean annual temperature. Wilson *et al.* (2007) showed uphill shifts of approximately 293 m in butterfly communities in the Sierra de Guadarrama of central Spain between 1967-1973 and 2004-2005 as a result of climate warming. Climate change may also weaken the association between climatic and habitat suitability. Franco *et al.* (2006) concluded the importance of climate warming and habitat loss in driving local extinctions of northern species of butterflies in northern Great Britain over the past few decades.

Forest pests are also occurring outside historic infestation ranges and at intensities not previously observed. Some examples of forest pest species that have responded or are predicted to respond to climate change by altering distribution include the following.

- A major epidemic of the mountain pine beetle (*Dendroctonus ponderosae*) has been spreading northwards and upwards in altitude in western Canada (British Colombia and more recently, Alberta) for several years.
- Warmer temperatures have influenced the southern pine beetle (*D. frontalis*) resulting in range expansions in the United States.
- The pine processionary caterpillar (*Thaumetopoea pityocampa*) has significantly expanded its latitudinal and altitudinal distribution in Europe.
- The oak processionary caterpillar (*T. processionea*) has shifted its distribution north in Europe during the latter half of the 20th century.
- The European rust pathogen *Melampsora allii-populina* is likely to spread northwards with increased summer temperatures.
- The root rot pathogen *Phytophthora cinnamomi* is predicted to spread into colder regions of Europe and have increased severity with climate change scenarios of increased average temperatures.

The ability of a species to respond to global warming and expand its range will depend on a number of life history characteristics, making the possible responses quite variable among species. Bale *et al.* (2002) suggested that fast-growing, non-diapausing insect species or those not dependent on low temperature to induce diapause, will respond to warming by ex-panding their distribution whereas slow-grow-ing species which need low temperatures to induce diapause (*i.e.* boreal and mountain species in the northern hemisphere) will suffer range contractions.

Range-restricted species, particularly polar and montane species, show more severe range contractions than other groups and are considered most at risk of extinction due to recent climate change. Range shifts may be limited by factors such as day length or the presence of competitors, predators or parasitoids. For example, the range expansion of insects which are very host-specific (specialists) may be limited by the slower rate of spread of their host plant species.

Phenology

Phenology is the timing of seasonal activities of plants and animals such as flowering or breeding. Since it is in many cases temperature dependent, phenology can be expected to be influenced by climate change. It is one of the easiest impacts of climate change to monitor and is by far the most documented in this regard for a wide range of organisms from plants to vertebrates. Common activities to monitor include earlier breeding or first singing of birds, earlier arrival of migrant birds, earlier appearance of butterflies, earlier choruses and spawning in amphibians and earlier shooting and flowering of plants.

Evidence of phenological changes in numerous plant and animal species as a consequence of climate change is abundant and growing. In general, spring activities have occurred progressively earlier since the 1960s and has been documented on all but one continent and in all major oceans for all well-studied marine, freshwater, and terrestrial groups.

Where life cycle events are temperature-dependent, they may be expected to occur earlier and increased temperatures are likely to facilitate extended periods of activity at both ends of the season, provided there are no other constraints present. With increased temperatures, it is expected that insects will pass through their larval stages faster and become adults earlier. Therefore expected responses in insects could include an advance in the timing of larval and adult emergence and an increase in the length of the flight period. Members of the Order Lepidoptera again provide the best examples of such phenological changes. Changes in butterfly phenology have been reported from the UK where 26 of 35 species have advanced their first appearance. First appearance for 17 species in Spain has advanced by 1-7 weeks in just 15 years. Seventy percent of 23 butterfly species in California, USA have seen an advancement of first flight date of approximately eight days per decade.

Changes in phenology - early adult emergence and an early arrival of migratory species - have also been noted for aphids in the UK. Gordo and Sanz (2005) investigated climate impacts on four Mediterranean insect species (a butterfly, a bee, a fly and a beetle) and noted that all species exhibited changes in their first appear-ance date over the last 50 years which was correlated with increases in spring tem-perature.

Parmesan and Yohe (2003) estimated that more than half (59 percent) of 1598 species investigated exhibited measurable changes in their phenologies and/or distributions over the past 20-140 years. They also estimated a mean advancement of spring events by 2.3 days/decade based on the quantitative analyses of phenological responses for these species. Root *et al.* (2003), in a similar quantitative study, estimated an advancement of 5.1 days per decade. Parmesan (2007) investigated the discrepancy between these two estimates and noted that once the differences between the studies in selection criteria for incorporating data was accounted for, the two studies supported each other, with an overall spring advancement of 2.3-2.8 days/decade found in the resulting analysis. However, in this last study, latitude explained only 4 percent of overall variation of phenological changes even though it is strongly associated with the importance of warming trends. This last observation may relate to the importance of the change in climate relative to the natural amplitude of the climate variability.

INDIRECT IMPACTS ON COMMUNITY ECOLOGY

Climate change is expected to alter the relationships between pests, their environment and other species, such as natural enemies, competitors and mutualists, leading to changes in the structure and composition of natural communities. The observed and predicted changes on species abundance and in phenological patterns and distributions of individual species are likely to alter species interactions within communities. Since individual species will respond to climate change in different ways and at different temporal scales there is a good possibility that some highly evolved relationships will be impacted. Interactions that involve two or more trophic groups, such as plant-herbivore, plant-pollinator and host-parasitic interactions are likely to suffer the largest mismatch (Harrington, Woiwod and Sparks, 1999).

In a review of phenological changes of interacting species, Visser and Both (2005) noted that insects have advanced their phenology faster (early eggs hatching and early mi-gration return date) than their hosts (budburst and flowering). They have also advanced their period of peak abundance more than their predators (laying date and migration arrival of birds). For example, the disruption of synchrony between the winter moth (*Operophtera brumata*) hatching and bud burst of its host oak trees has in turn resulted in an asynchrony between the pest and one of its predators, the great tit (*Parus major*), which relies on the caterpillars to feed their young. Such climate-induced phenological changes are clearly resulting in a great deal of asynchrony between interacting species which will ultimately influence community structure, composition and diversity.

Distributional changes and range shifts interfere with community relationships as expanding species will begin to interact with other species in new environments with which previous interaction may have been limited or non-existent. The altitudinal spread of the pine processionary caterpillar (*Thaumetopoea pityocampa*) in the Sierra Nevada mountains of southeastern Spain for example, has resulted in the pest encountering a new host tree, the endemic Scots pine (*Pinus sylvestris* var. *nevadensis*). Increased attacks by *T. pityocampa* could have deleterious effects on this endemic mountain tree species.

Species capable of responding to climate change by increasing their range will also benefit from the lack of competitors and natural enemies in their new environment. Species expansions may not be promptly followed by that of its enemies, as in the case of the pine processionary moth, and in some cases, the synchronization between host and enemy or parasitoid may not be maintained under new temperature conditions, as is the case with the winter moth.

Some pathogens may benefit from the improved survival and spread of their insect vectors. For example, the vectors of Dutch elm disease (*Ophiostoma novo-ulmi*), *Scolytus scolytus* and *S. multistriatus* may be more active during periods of elevated temperature which would ultimately result in increased spread of the fungus.

INDIRECT IMPACTS OF CLIMATE CHANGE ON HOST TREES

Changes in temperature, precipitation, atmospheric CO_2 concentrations and other climatic factors can alter tree physiology in ways that affect their resistance to herbivores and pathogens.

DROUGHT

Drought is one of the most important climate-related events through which rapid ecosystem changes can occur as it affects the very survival of existing tree populations. Long-term drought can result in reduced tree growth and health thereby increasing their susceptibility to insect pests and pathogens. A number of insect pests and diseases are associated with stressed trees, such as *Agrilus* beetles and the common and widespread *Armillaria* species which have been linked to oak decline. Others are limited by host defences in healthy trees, such as the European spruce bark beetle (*Ips typographus*).

Drought can also elicit changes in plant and tree physiology which will impact pest disturbance dynamics. Leaves may change colour or become thicker or waxier which could affect their palatability to insects. The concentration of a variety of secondary plant compounds tends to increase under drought stress which would also lead to changes in the attraction of plants to insect pests. Moderate drought however may actually increase production of defence compounds in plants and trees possibly providing increased protection against pests.

Sugar concentrations in foliage can increase under drought conditions making it more palatable to herbivores and therefore resulting in increased levels of damage (Harrington, Fleming and Woiwod, 2001). Increases in the sugar content in drought-stressed balsam fir for example have been known to stimulate the feeding of certain stages of spruce budworm (*Choristoneura fumiferana*) and accelerate their growth. Another advantage for forest pests is the increased temperature of drought-stressed trees, which can be 2-4 °C warmer, which can benefit the fecundity and survival of insects for example.

The impacts of such changes to host tree physiology and susceptibility provoke different responses from pest species. Rouault *et al.* (2006) investigated the impacts of drought and high temperatures on forest insects and noted that woodborers were positively influenced by the high temperatures which increased their development rates and the prolonged water stress that lowered host tree resistance while defoliators benefited from the increased nitrogen in plant tissues linked to moderate or intermittent water stress.

The large natural spatial and temporal variability in forest processes makes it difficult to positively relate drought-related tree mortality to a greater incidence of pest or fungal pathogen damage. In a recent study on the impact of a large-scale, multi-annual drought on the growth and mortality of aspen (*Populus tremuloides*) in Canada, Hogg, Brandt and Michaelian (2008) could not find a significant relationship between drought severity and either insect defoliation or wood borer infestations. Drought severity was, however strongly related to total growth loss and mortality, but the delay of mortality into the years following the drought suggested that secondary agents may have been involved in the process.

ELEVATED LEVELS OF ATMOSPHERIC CARBON DIOXIDE

Higher atmospheric CO_2 levels result in improved growth rates and water use efficiency of plants and trees. This increased productivity leads to lower nitrogen concentrations in trees and plants as carbon-nitrogen (C: N) ratios rise and thus reduces the nutritional value of vegetation to insects. In response insects may increase their feeding (and consequently tree damage) in an attempt to compensate for the reduced quality and gain the necessary nitrogen. In many cases the increased feeding does enable the insect to meet its nutritional needs but most often it does not and results in poor performance, reduced growth rates and increased mortality. Such an effect, however, is not consistently observed, and increased growth due to enhanced CO_2 may in fact more than compensate for the defoliation in some cases.

Elevated CO_2 levels can also result in changed plant structure such as increased leaf area and thickness, greater numbers of leaves, higher total leaf area per plant, and larger diameter stems and branches. An increase in defensive chemicals may also result under such conditions (van Asch and Visser, 2007). Either of these changes to host physiology would influence palatability to insects, though the impacts on pests differ by species. For example, under increased CO_2 levels the winter moth (*Operophtera brumata*) consumes more oak (*Quercus robur*) leaves due to a reduction in leaf toughness, while the gypsy moth (*Lymantria dispar*) exhibits normal pupation weight but requires a longer time to develop as a result of an increase in tannin concentrations (van Asch and Visser, 2007).

Nitrogen Deposition

Anthropogenic emissions of nitrogenous air pollutants and their subsequent deposition are part of the larger phenomenon of global changes and can also have impacts on forest health. Increased nitrogen levels in the leaves of trees and plants can result in an immediate increase in the incidence of a number of pathogens (Burdon, Thrall and Ericson, 2006).

For example, infections caused by the common pathogen, *Valdensia heterodoxa*, on *Vaccinium myrtillus*, a dominant understorey plant in boreal coniferous forests, are boosted by increased nitrogen availability resulting in premature leaf-shedding (Burdon, Thrall and Ericson, 2006). This defoliation subsequently promoted a shift from *V. myrtillus* dominance to grass dominance thereby affecting community structure.

Extreme Events

Besides drought, climate change may affect the frequency and intensity of other extreme climate-related events, with subsequent impacts on forest health.

Direct damage to trees or alterations in the ecosystem may increase their susceptibility to pest outbreaks. Windstorms and lightning strikes can damage trees and allow entry of pathogens and secondary insect pests as well as causing mechanical breakdown in normal physiological function.

FOREST PEST SPECIES INFLUENCED BY CLIMATE CHANGE

Some examples of forest insect pests, diseases and other pests which have been impacted or are predicted to be impacted by climate change are presented below. Information on non-forest pests is also provided to enable a better understanding of the potential impacts of climate change on forest health.

HEMIPTERA

Aphids (Aphididae)

With short generation times and low developmental threshold temperatures, aphids are a group of insects that can be expected to be strongly influenced by environmental and climatic changes. In general, it has been predicted that aphids will appear at least eight days earlier in the spring within 50 years, though the rate of advance will vary depending on location and species. This could potentially result in greater damage to host plants depending on the phenology of host plants and natural enemies.

Zhou *et al.* (1995), for example, investigated the timing of migration in Great Britain for five aphid species (*Brachycaudus helichrysi, Elatobium abietinum, Metopolophium dirhodum, Myzus persicae, Sitobion avenae*) over a period of almost 30 years and concluded that temperature, especially winter temperature, is the dominant factor affecting aphid phenology for all species. They found that a one degree Celsius increase in average winter temperature advanced the migration phenology by 4-19 days depending on species.

Elatobium abietinum (Walker) (Aphididae)

The green spruce aphid (*Elatobium abietinum*) is also believed likely to benefit from the increase in winter survival, leading to more intense and frequent defoliation of host spruce trees (*Picea* spp.). This aphid is native to Europe but has also been reported in both North and South America.

Infestations in Great Britain have resulted in large losses of spruce foliage and height both during the active infestation and in subsequent years. Westgarth-Smith *et al.* (2007) showed that warm weather associated with a positive North Atlantic Oscillation (NAO) index caused spring migration of *E. abietinum* to start earlier, last longer and contain more aphids. Positive NAO values correspond to warmer atmospheric conditions over Great Britain. Since global warming is believed to increase NAO variability, shifting the system to more positive values, this will most likely lead to further increases in aphid activity and more damage to spruce trees and forests in the area.

COLEOPTERA

Agrilus Pannonicus

A number of Buprestid beetles of the genus *Agrilus* have been linked to oak decline. Incidences of these species have increased worldwide (both in their countries of origin and by international movement) and their impacts are being linked to host tree stress potentially caused by climate change. For example, *Agrilus pannonicus* (=*A. biguttatus* (Fabricius)) has recently been associated with a European oak decline throughout its natural range and has increased in incidence in several countries including France, Germany, Hungary, Poland and the Netherlands, and the UK where it is believed to be contributing to oak decline. Infestations can result in extensive tree mortality which, combined with other factors involved in the decline, can drastically alter the species composition of oak forests.

Dendroctonus Frontalis Zimmermann

Dendroctonus frontalis is considered to be one of the most damaging species of bark beetles in Central America and southern areas of North America. It is a major pest of pines and has a wide distribution occurring from Pennsylvania in the United States south to Mexico and Central America. Populations can build rapidly to outbreak proportions and large numbers of trees are killed. Initial attacks are generally on weakened trees however *D. frontalis* is capable of killing otherwise healthy trees. This beetle kills trees by a combination of two factors: girdling during construction of egg galleries; and the introduction of blue stain fungi of the genus *Ophiostoma.* Because of their short generation time, high dispersal abilities and broad distribution of suitable host trees, the southern pine beetle has the potential to respond quickly and dramatically to any changes in climate.

In October 1998, Hurricane Mitch hit Central America, causing floods and mudslides that ravaged local communities, forests and infrastructure. In the years that followed an unprecedented regionwide outbreak of pine bark beetles, mainly *D. frontalis* in association with other *Dendroctonus* and *Ips* species, destroyed over 100 000 ha of pine forest. As most of the standing dead and felled trees were left on site, fuel loads were drastically increased thereby resulting in extensive wildfires. With climate change expected to increase the frequency and severity of extreme events such as hurricanes, the potential for future devastating impacts on forests from both the initial disturbance and its cascading effects (*i.e.* other disturbances such as pest outbreaks and fire) is quite high.

Warmer temperatures attributed to climate change have also influenced the southern pine beetle resulting in range expansions in the United States. Laboratory measurements and published records of mortality in wild populations indicate that a temperature of -16 °C or less result in almost 100 percent mortality of the pest, thereby limiting its distribution in its current northern range. It was predicted that an increase in temperature of 3 °C would enable outbreaks to occur approximately 178 km farther north than in historical times (Ungerer, Ayres and Lombardero, 1999). Recent outbreaks of the southern pine beetle in northern and high-altitude ecosystems, where they were previously rare or absent, have been attributed to a warming trend of 3.3 °C in minimum winter air temperatures in the southeastern US from 1960-2004. This northern expansion is about as predicted by Ungerer, Ayres and Lombardero (1999).

The southern pine beetle has also possibly adapted ways to increase survival in cooler climes. Tran *et al.* (2007) showed, through field and laboratory studies of a northern population, that prepupae were more cold tolerant (by more than 3 °C) than pupae, adults and feeding larvae, and that the winter life stage structure was strongly biased toward this most cold-tolerant life stage. This tendency to overwinter in a cold tolerant life stage could be a coincidence however rather than a true adaptation.

Dendroctonus Ponderosae

The mountain pine beetle (*Dendroctonus ponderosae*) is the most destructive pest of mature pines in North America, particularly lodgepole pine (*Pinus contorta*). In the western United States, outbreaks have been increasing in area after several years of drought (Tkacz, Moody and Villa Castillo, 2007). A major epidemic of this pest has also been ongoing in western Canada (British Columbia (BC), and more recently, Alberta) for several years and even with large-scale efforts to mitigate the impacts of the pest, millions of trees have been killed. A record of over 10 million hectares of pines were recorded as infested during 2007 aerial overview surveys in BC, with 860 973 ha of this located in provincial parks and protected areas. It has been predicted that if the beetle continues to spread at its current rate as much as 80 per cent of mature pine in BC will be dead by 2013. The large numbers of dead and dying trees have also increased the risk of wildfires.

The problem has been exacerbated by successive years of mild winters, resulting in decreases of mortality of overwintering stages and generation time. Their life cycle is generally completed in one year; warmer temperatures can result in two generations per year while cooler ones may results in one generation every two years (Amman, McGregor and Dolph, 1990). Drought conditions associated with warmer temperatures have also weakened the trees and increased their susceptibility to the beetles. Warmer temperatures have thus opened up previously climatically unsuitable mature pine stands to the pest.

Dendroctonus Rufipennis

Dendroctonus rufipennis is a North American pest of spruce, particularly white spruce (*Picea glauca*) and black spruce (*P. mariana*) in the north, Engelmann spruce (*P. engelmannii*) and sitka spruce (*P. sitchensis*) in the west, and red spruce (*P. rubens*) in the east. It tends to attack weakened or windthrown trees and outbreaks are mostly linked to predisposing factors. As a result it can be expected that the impacts of climate change on trees and forests could enhance spruce beetle outbreaks.

In fact, Hebertson and Jenkins (2008) investigated the impact of climate on spruce beetle outbreaks in Utah and Colorado, USA between 1905 and 1996 and found that historic outbreak years in the intermountain region were related to generally warm fall and winter temperatures and drought conditions. Similarly outbreaks in both Canada (Yukon Territory) and the US (Alaska) appear to be related to extremely high summer temperatures which influenced spruce beetle population size through a combination of increased overwinter survival, a halving of the maturation time from two years to one year, and regional drought-induced stress of mature host trees.

HYMENOPTERA

Cephalcia arvensis Panzer

The spruce webspinning sawfly is monophagous on spruce (*Picea*) and endemic to the spruce range in Eurasia, where outbreaks have been seldom recorded. From 1985-1992 however there was a sudden outbreak of the sawfly in the Southern Alps during which populations developed an annual life cycle and grew exponentially, causing repeated defoliations resulting in extensive tree death. *Cephalcia* species generally show low fecundity and have an extended diapause of a few years that is stimulated by low temperatures at pupation time. The outbreak corresponded to a period of high temperatures and low precipitation and severe water stress for the host trees. As a result, the insect was able to adapt to the new climate resulting in lower mortality, faster development and higher feeding rates of the sawfly. In addition, the sudden increase in population density was not quickly followed by that of natural enemies, thereby allowing for unlimited population growth.

Neodiprion Sertifer

The European pine sawfly *Neodiprion sertifer* is an important pest species on pines in Europe, northern Asia, Japan and North America where it was introduced. It is one of the most serious defoliators of Scots pine (*Pinus sylvestris*) forests in northern Europe. Virtanen *et al.* (1996) suggested that outbreaks of the sawfly on Scots pine in eastern and northern Finland are prevented by low winter temperatures which kill eggs, and predicted that outbreaks would become more common with winter warming. A high variation in freezing avoidance of eggs was also noted which would allow *N. sertifer* to adapt to predicted climate change and spread its distribution northwards.

ASCOMYCOTA

Mycosphaerella pini Rostrup

Mycosphaerella pini is a fungus that infects and kills the needles of *Pinus* spp. resulting in significant defoliation, stunted growth and eventually death of host trees although susceptibility among pine species does vary. Native to temperate forests of the northern hemisphere, it is a major pest of pine plantations in the Southern Hemisphere, where both the host and the pathogen have been introduced. The widely planted *P. radiata* is particularly susceptible and many forests planted with this species in the Southern Hemisphere, particularly in East Africa, New Zealand and Chile, have been devastated by this needle blight. This pathogen has forced managers in some areas to abandon the planting of *P. radiata* and depend more on other tree species.

In its native range *M. pini* normally causes little damage, but since the late 1990s it has been causing extensive defoliation and mortality in young plantations of lodgepole pine (*Pinus contorta* var. *latifolia*) in northwestern British Columbia, Canada (Woods, Coates and Hamann, 2005). Mortality of mature lodgepole pines has been observed in mixed-species stands, where scattered pine represents only a small proportion of stand composition; this represents a globally unprecedented occurrence for *M. pini* (Woods, Coates and Hamann, 2005). The current epidemic coincides with a prolonged period of increased frequency of warm rain events throughout the mid-to-late 1990s allowing for the rapid spread and increased rates of infection. Unlike many other pests, changes in precipitation patterns may be more important than changes in temperature for predicting the spread and impact of *M. pini*.

LEPIDOPTERA

Warmer temperatures have been linked to increasing populations of forest Lepidoptera species.

Butterflies

While only a few species of butterflies are considered to be serious forest pests, some of the best, and most researched, examples of the impacts of climate change on insect distributions and phenology have been butterflies. The geographic ranges of many species have shifted northward and upwards in elevation associated with climate warming, leading to increases in species richness at high latitudes and elevations and in some cases possible local extinction at lower altitudes.

Range expansions in butterflies have been well documented and changes in butterfly phenology have also been reported. In the UK, species have been advancing their flight periods by approximately 2-10 days for every 1 ºC increase in temperature. Similar changes in phenology as a response to warming has been noted in Spain where butterflies have advanced their first appearance by 1-7 weeks in 15 years and in California, USA which has seen an advancement of approximately eight days per decade .

Some species-specific examples of the influence of climate change include the following.

- The African Monarch butterfly (*Danaus chrysippus*) has spread northward, establishing its first population in southern Spain in 1980 followed by the establishment of multiple populations along the east coast of Spain.
- Warmer temperatures have increased survival and facilitated a latitudinal and altitudinal range expansion of *Atalopedes campestris* in the western USA.
- Edith's checkerspot butterfly (*Euphydryas editha*) has shifted its distribution northwards and also upwards in altitude in North America. Populations at the northern edge of the species range in Canada and also at higher altitudes within the main range have experienced increased survival whereas populations at the southern edge in Mexico have declined.
- In Europe the speckled wood butterfly (*Parage aegeriae*) has increased its range northward beyond its original, primary host.
- The black-veined white butterfly, *Aporia crataegi*, has expanded its altitudinal range in the mountains of the Sierra de Guadarrama of central Spain resulting in local population extinctions at lower warmer altitudes. While climate is becoming less of a limiting factor in its distribution at higher altitudes, it is however limited by the absence of host plants.

Choristoneura Fumiferana

The spruce budworm, *Choristoneura fumiferana*, is a major defoliator of coniferous forests across North America. Balsam fir (*Abies balsamea*) is the preferred host but they readily attack white, red and black spruce (*Picea glauca*,

P. rubens, P. mariana respectively) and may even be feed on tamarack (*Larix* spp.) and hemlock (*Tsuga* spp.). Outbreaks of this budworm can persist for 5-15 years with periods of 20-60 years in between. In eastern Canada the period of population cycle has averaged 35 years over the last 270 years. During uncontrolled outbreaks they can kill almost all trees in dense, mature stands of fir. Climatic influences on life history traits are considered a major factor in restricting outbreaks and as a result, a changing climate is expected to impact the severity, frequency, and spatial distribution of spruce budworm outbreaks. The success of the insects in establishing feeding sites in the spring depends on initial egg weights and synchrony of their development with that of buds of their hosts which is strongly influenced by climatic factors. This synchronization is critical in initiating outbreaks and thus determining the intensity of damage. However the spruce budworm is able to tolerate some asynchrony between spring emergence and vegetative shoot development as the second instars have adapted morphologically and behaviourally allowing them to mine needles.

In parts of its range, particularly at northern extremes, temperature can also influence the duration of outbreaks as collapses are often associated with the loss of suitable foliage often as a result of late spring frosts. Normal collapse of the outbreak in the core range of host trees is associated with mortality caused by natural enemies late in the larval stage. Natural enemies of the spruce budworm, *C. fumiferana*, are less effective at higher temperatures and therefore climatic factors have the potential to enable massive outbreaks of this pest providing there is suitable availability of host trees.

Epirrita Autumnata

Epirrita autumnata is a holarctic species that has been expanding its outbreak range in some areas. In the Nordic countries of Europe, *Epirrita autumnata* outbreak cycles are typically most prevalent in northernmost and continental birch forests but during the past 15-20 years it has expanded into the coldest, most continental areas previously protected by extreme winter temperatures. This pest overwinters in the egg stage and therefore the level of egg survival is dictated by minimum winter temperatures. Virtanen, Neuvonen and Nikula (1998) investigated the relationship between *E. autumnata* egg survival and minimum winter temperatures in northernmost Finland. They predicted that climate warming would result in a two-third reduction of the area of forests with winter temperatures cold enough to keep *E. autumnata* populations low by the middle of the next century. A rise in winter temperatures therefore will likely increase the area of forest susceptible to damage by the autumnal moth.

Lymantria Dispar Linnaeus

The gypsy moth, *Lymantria dispar*, is a significant defoliator of a wide range of broadleaf and even conifer trees. While low population levels can exist for many years without causing significant damage, severe outbreaks can occur resulting in severe defoliation, growth loss, dieback and sometimes tree mortality. Two strains of gypsy moth exist – the Asian strain, of which the female is capable of flight; and the European strain, of which the female is flightless. The Asian strain is native to southern Europe, northern Africa, central and southern Asia, and Japan and has been introduced into Germany and other European countries where it readily hybridizes with the European strain. It has also been introduced but has not established in Canada, the US and the UK (London). The European strain is found in temperate forests throughout Western Europe and has been introduced into Canada and the US. The gypsy moth is considered a significant pest in both its native and introduced ranges.

There has also been a noted increase in outbreaks in areas previously unaffected by this pest such as the Channel Islands (Jersey) and new areas in the UK (Aylesbury, Buckinghamshire). In Canada the spread of the gypsy moth has so far been prevented by climatic barriers and host plant availability as well as by aggressive eradication of incipient populations. However it is predicted that the gypsy moth will be able to extend its range in North America as a result of higher overwinter survival of egg stages because of milder winters and higher accumulation of day degrees for larval development. Similar predictions have been made for other areas. For example, Pitt, Régnière and Worner (2007) noted an increase in the probability of establishment of the gypsy moth in New Zealand, particularly in the South Island.

Increasing atmospheric concentrations of CO_2 may also influence the severity of gypsy moth outbreaks. Larval performance on host plants grown under elevated CO_2 varies depending on the species, being reduced on some hosts such as aspen and increased on others, such as oak.

Lymantria Monacha

Lymantria monacha is a major pest of broadleaved and coniferous trees in Europe and Asia. Defoliation by nun moth larvae can kill host trees especially conifers and has caused extensive losses despite intervention with biological and chemical insecticides. In parts of Europe, the occurrence of outbreaks has increased possibly as a result of the establishment of extensive pine plantations in poor quality areas or as a result of a changing climate. It has been predicted that nun moth will spread northwards in Europe because of higher accumulated day degrees and improved overwinter survival. Using modelling software, Vanhanen *et al.* (2007) predicted that climate warming would shift the northern boundary of distribution north by approximately 500-700 km and the southern edge of the range would retract northwards by 100-900 km.

Operophtera brumata Linnaeus

Operophtera brumata is distributed throughout Europe, North Africa, Japan and Siberia and has also been introduced into Canada and the US. It feeds on a variety of deciduous trees and shrubs including apricot, cherry, apple, plum, blueberry, crab-apple, sweet chestnut, red currant and black currant, oaks, maples, basswood and white elm. Climate change is impacting the spread of the winter moth. In the Nordic countries of Europe, Jepsen *et al.* (2008) noted that *O. brumata* had been climatically restricted to more southern and near-coastal locations in the regions but warmer temperatures has resulted in expansions in its outbreak area further northeast. While increased temperatures appear to assist the winter moth expand its distribution, it appears that they do not have the same impact on its natural enemies which may allow populations of this pest to grow unchecked.

Climate change has affected the phenology of many species in different ways. In the Netherlands over the past 25 years, early spring temperatures have increased while winter temperatures have not. As a result a climate change induced asynchrony has occurred between winter moths and their host, pedunculate oak, *Quercus robur*, with eggs hatching before bud burst . Such a situation leaves no food for the larvae resulting in starvation and death. This also has implications for other species that depend on the larvae for food such as the great tit (*Parus major*) which feed *O. brumata* caterpillars to their young. While both egg hatch and bud burst have advanced over the last 25 years, egg hatch has advanced much more leading to a decrease in synchrony from a few days to almost 2 weeks. However others have noted that, while warmer temperatures have lead to earlier egg hatch, the autumnal pupal diapause of the winter moth is prolonged at higher temperatures thereby counteracting the impact and resulting in a life cycle that is not shortened overall. Differences between observations of synchronicity between moth egg hatch and host bud burst may result from regional, intra-specific differences.

Thaumetopoea pityocampa

The pine processionary caterpillar, *Thaumetopoea pityocampa*, is considered one of the most important pests of pine forests in the Mediterranean region (EPPO/CABI, 1997). It is a tent-making oligophagous caterpillar that feeds gregariously and defoliates various species of pine and cedar. The life cycle of the pine processionary caterpillar is typically annual but may extend over two years at high altitudes or in northern latitudes. At northern latitudes and at higher altitudes, adults emerge earlier. Climate change is having clear impacts on the distribution of this important forest pest. Battisti *et al.* (2005) reported a latitudinal expansion in north-central France of 87 km northwards from 1972-2004 and an altitudinal shift of 110-230 m upwards in the Alps of northern Italy from 1975-2004 and attributed the expansions to reduced frequency of late frosts, which increases survival of overwintering larvae, as a result of a warming trend over the past three decades. In the last ten years the pine processionary caterpillar has spread almost 56 km northward in France. During the summer of 2003, the warmest summer in Europe in the last 500 years, *T. pityocampa* exhibited an unprecedented expansion to high elevation pine stands in the Italian Alps, increasing its altitudinal range limit by one third of the total altitudinal expansion over the previous three decades. This unusual and fast spread has been attributed to increased nocturnal dispersal of females during the unusually warm night temperatures. The gradual warming of the region has allowed the pest to maintain its presence at this altitude because of increased larval survival.

In the Sierra Nevada mountains of southeastern Spain, *T. pityocampa* has expanded to higher elevations over the last 20 years as a result of increasing mean temperature. Relict populations of Scots pine (*Pinus sylvestris* var. *nevadensis*) occurring within this newly expanded range of the caterpillar are being increasingly attacked, particularly in warmer years. This range expansion caused by climate change has potentially devastating consequences for this endemic mountain species which is likely to suffer from the direct effects of climate change as well. Given that the present distribution of the *T. pityocampa* is not constrained by the distribution of its hosts, that warmer winters will increase winter larval feeding activity, that the probability of lower lethal temperature will decrease, it can be expected that the improved survival and spread into previously hostile environments will continue.

Thaumetopoea Processionea

Native to central and southern Europe, *Thaumetopoea processionea* is a major defoliating pest of oak. Since the late 20th century it has been expanding its range northwards and is now firmly established in Belgium, Denmark, northern France, and the Netherlands and has been reported from southern Sweden and the UK. It is believed that the northward progression of the oak processionary moth is due to improved synchrony of egg hatch and reduction of late frosts as a result of warmer temperatures.

Zeiraphera Diniana Guenée

The larch bud moth, *Zeiraphera diniana*, is a European pest that has been defoliating large areas of larch forests in the Alps every 8-10 years for centuries. It has an annual life cycle, overwintering as an egg on the larch branches and feeding on the needles as soon as the bud breaks. As such synchrony between egg hatch and bud burst is critical. Increased temperatures associated with climate change have affected this relationship leading to asynchrony and reduced incidences of the moth in Switzerland. It has been reported that abnormally high temperatures result in unusually high egg mortality.

BASIDIOMYCOTA

Armillaria Species

Armillaria species are common worldwide pathogens of trees, woody shrubs and herbaceous plants that can cause wood decay, growth reduction and even mortality, particularly in trees stressed by other factors, or in young trees planted on sites from which infected hosts have been removed. *Armillaria* species can become more aggressive and damaging when elevated temperatures cause drought stress thereby reducing tree defences. Tree physiological condition in general may be an important factor in controlling the impacts of *Armillaria* species, and climate change may affect their epidemiology.

OOMYCOTA

Phytophthora Cinnamomi

Phytophthora cinnamomi is considered one of the most widely distributed and destructive forest pathogens. It has a wide host range infesting over 1000 species resulting in root rot and cankering. The native range is unknown but it is believed to be from Southeast Asia and southern Africa. Currently the pathogen can be found in most temperate and subtropical areas in the world in Africa, Asia and the Pacific, Europe, Latin America and the Caribbean, Near East, and North America. In most countries it is only known in nurseries but in Europe (France, Italy, Spain, Portugal) it has observed in natural environments. Temperature, moisture and pH all influence the growth and reproduction of the fungus. In a study on the impacts of climate warming on *P. cinnamomi*, Bergot *et al.* (2004) predicted a potential range expansion of the disease in Europe of one to a few hundred kilometres eastward from the Atlantic coast within one century.

PHYTOPHTHORA RAMORUM WERRES

Phytophthora ramorum causes a very serious disease called sudden oak death which causes extensive mortality of tanoak and oaks. It is also associated with disease on ornamental plants and other broadleaf and conifer trees. This pathogen is a significant problem in both North American and European forests and nurseries. The geographic origin of *P. ramorum* is unknown; it is believed that it has been introduced independently to Europe and North America from an unidentified third country. The pathogen likely disperses through a variety of means. Sporangia may be dispersed locally by rain splash, wind-driven rain, irrigation or ground water, soil and soil litter. Bark and ambrosia beetles are commonly found on infected trees but their potential role of vectors has not yet been investigated. Consequently, changes in climate, precipitation and temperature in particular, will likely produce more optimal conditions for the pathogen resulting in an increase in disease occurrence.

NEMATODES

In general there is a close correlation between soil temperatures and the distributions of some plant-parasitic species of nematode. For example,

Meloidogyne incognita, previously deemed limited to the Mediterranean area, was recently found in the Netherlands. It is also believed that a one degree Celsius rise in temperature would allow *Longidorus caespiticola* to become established further north in Great Britain.

Bursaphelenchus Xylophilus

The pine wilt nematode, *Bursaphelenchus xylophilus,* is the causal agent of pine wilt disease and is spread by *Monochamus* beetles. Native to North America where it is not considered a serious pest, the nematode is a major threat to Asian and European pine forests and has resulted in extensive tree mortality in countries where it has been introduced. Changes in both temperature and precipitation are likely to impact the spread of the nematode and the severity of damage caused by the disease. Pine wilt disease is most prevalent in warm climates as the nematode completes its life cycle in 12, 6 and 3 days at 15, 20 and 30°C, respectively. High temperatures and low precipitation in summer cause accelerated damage through their impacts on vector activity, propagation of the nematode and water stress on trees. In Japan, while annual tree losses to the disease have gradually decreased, infestations have spread into northern areas and into forests at higher elevations as a result of increased temperatures.

2

Forestry Resources and Climate Change

GREENHOUSE GASES AND GLOBAL WARMING

Human activities have resulted in the alteration of the composition of our atmosphere triggering change in the Earth's climate. The world's population has grown at an alarming rate with a corresponding increase in demand for natural resources, energy, food, and goods. As a consequence of increase in consumption, vast quantities of gases and effluents are discharged that change the composition of the atmosphere and its capacity to regulate its temperature.

The rise in the global temperature is caused by the accumulation of the so-called "greenhouse gases", namely, carbon dioxide (CO_2), methane, nitrous oxide, and chlorofluorocarbons. Energy received from the sun is absorbed as short wavelength radiation and is eventually returned to space as long wavelength infrared radiation.

Greenhouse gases absorb the infrared radiation, trapping it in the atmosphere in the form of heat energy. By increasing the atmosphere's ability to absorb infrared energy, the greenhouse gases are disturbing the way the climate maintains the balance between incoming and outgoing energy. This increases the probability of occurrence of unseen and unpredictable events across the planet.

There has been an increase of average global surface temperature by 0.6 ^{0}C during the last 100 years as a consequence of human activities, such as deforestation and burning of fossil fuels. The Intergovernmental Panel on Climate Change (IPCC) stated in its Third Assessment Report that the globally averaged surface temperature is projected to increase by 1.4 °C to 5.8 °C from 1990 to 2100 under business-as-usual, and sea levels by 9 cm to 88 cm over the same period. If nothing is done to prevent or limit these changes, they will have major consequences for the ecosystem.

EFFECTS OF CLIMATE CHANGE

The effects of climate change are manifest in several ways Hitz S., Smith J. (2004). Estimating global impacts from climate change. Global Environmental Change 14, 201-218. Globally, precipitation is on the rise. In the Northern Hemisphere, precipitation has increased by 0.5% to 1.0% per decade whereas the increase in tropical countries has been 0.2% to 0.3% per decade. The changes in the global climate have caused a reduction of the snow pack in northern latitudes, a melting of mountain glaciers, a thawing of the Arctic permafrost, and a shrinking of the polar ice caps. The average sea level of the world's oceans has risen 10 cm to 20 cm in the last century. UNEP models predict an increase in global mean sea level of between 13 cm and 94 cm by the year 2100; however, other models predict a sea level rise by 54cm to 68 cm by 2100.

These consequences will include geographic shifts in the occurrence of different species and/or the extinction of species. Changes in rainfall patterns will put pressure on water resources in many regions, which will, in turn, affect both irrigation and drinking water supplies. Extreme weather events and floods will become more frequent with their increasing economic costs and human suffering. River deltas that are presently being farmed and estuaries that are now important wildlife habitats will be flooded and become saline, making them unsuitable for these uses. Warm seasons will become dryer in the interior of most mid-latitude continents, increasing the frequency of droughts and land degradation. This will be particularly serious for areas where land degradation, desertification and droughts are already severe. Additionally, tropical diseases will extend beyond their present geographic range.

Developing countries would bear the maximum brunt of global warming and climate change. The socioeconomic impact of climate change in developing countries will be significant. Rising temperatures and irregular precipitation patterns will impact negatively on agricultural crop yields, food security, and health issues related to malnutrition. An increased incidence and intensity of violent storms and monsoons will produce more flooding which, in turn, will cause greater damage to infrastructure and an increase in the incidence of vector-borne (*e.g.*, malaria) and water-borne (*e.g.* cholera) diseases. The least privileged in society are also the least equipped to adapt to climate change. Poverty, limited infrastructure, poor access to technology, inadequate education, and limited management skills; combine to frustrate the ability of developing countries to respond to the challenges.

India's Contribution to Climate Change

Developing countries continue to use technology that though cheaper, is both outdated and harmful for the environment. Their less efficient use of fuels and other resources is from both a lack of state-of-art technology and proportionally higher use of coal and biomass, which emit more of GHGs per unit of energy than do petroleum products and natural gas. This further increases their vulnerability to global as well as local environmental and climatic changes. Though the developed countries contribute a major chunk to the global warming, still they are not as vulnerable as the developing countries. The total carbon dioxide-equivalent emissions from India are estimated to be 1,001,352 Gg (1000 Mt), which is about three percent of the total global carbon dioxide-equivalent emissions. Based on this, the per capita carbon dioxide-equivalent emissions for 1990 are estimated to be 1.194 tonnes or 325 kg of carbon. In comparison, the per capita emissions for Japan and the US are 2400 and 5400 kg of carbon respectively in 1990. India's CO_2 emissions are far below the developed countries but it will be one of the nations which could possibly be seriously affected by global warming and the resulting climate change, partly due to its high population density and partly due to its long coastline.

Forests and Climate Change

The burning of fossil fuels is the most important source of greenhouse gases. Coal and petroleum are the principal fuels for power generation, industry, and transportation, accounting for about 75 percent of all emissions. As the global economy has grown, there has been a dramatic increase in the consumption of fossil fuels, witnessed most dramatically in the virtual revolution of the transportation sector. Cars and trucks have increased from a mere handful a hundred years ago to the tens of millions of vehicles currently on the road. Airline travel has also gone up to thousands of flights per day. All of this change has been accompanied by a skyrocketing consumption of fossil fuels. Also, energy needs for heating and domestic cooking have not only increased with the growth in population, but also with the changes in technology. There has been a shift away from biomass fuels to more convenient, less expensive petroleum-based sources.

In addition to the burning of fossil fuels, other important sources of greenhouse gas emissions are activities related to land use, primarily tropical deforestation and forest fires. Greenhouse gases from deforestation are mostly carbon dioxide with lesser amounts of methane and carbon monoxide. Tropical deforestation is one of the most critical environmental problems facing the developing countries today in terms of its long-term and catastrophic impact on biodiversity, loss of economic opportunities, creation of social problems, and contribution to global climate change. Forests are both a source and a sink of carbon.

Thus, managing forests for carbon storage will help to absorb atmospheric carbon dioxide. Like agricultural crops, forests sequester carbon from the atmosphere as a result of photosynthesis. As trees have a much longer lifespan, they act as long-term reservoirs that lock up the carbon for decades, even centuries, in the form of cellulose and lignin.

Forestry and Kyoto Protocol

Forestry has assumed greater significance after the inclusion of afforestation and reforestation activities under the Kyoto Protocol of the United Nations Framework Convention on Climate Change (UNFCCC). UNFCCC is an international environmental treaty formulated at the United Nations Conference on Environment and Development (UNCED), informally known as the Earth Summit, held in Rio de Janeiro in 1992. The treaty aimed at reducing emissions of greenhouse gases, pursuant to its supporters' belief in the global warming hypothesis. The treaty as originally framed set no mandatory limits on greenhouse gas emissions for individual nations and contained no enforcement provisions; it is therefore considered legally non-binding.

The Kyoto Protocol was adopted in December 1997 and came into force in February 2005. The objective of this agreement is to stabilize greenhouse gas concentrations in the atmosphere at a level that would prevent dangerous anthropogenic interference with the climate system. Countries, which ratify this protocol, commit to reducing their emissions of carbon dioxide and five other greenhouse gases, or engage in emissions trading if they maintain or increase emissions of these gases. The industrialized countries are required to reduce their collective emissions by 5.2% compared to the year 1990. In addition to this the industrialized countries are also required to consider ways to minimize adverse affects of their greenhouse gas emissions on developing countries. The developing countries are not required by the Kyoto Protocol to reduce their greenhouse gas emissions, as the per-capita emission rates of the developing countries are a tiny fraction of those in the developed world. However, the capacity of the developing countries must be strengthened to enhance their contribution to fighting climate change. Capacity building and technology transfer needs to be a central part of the global approach to combating climate change.

Carbon mitigation in developing countries remains an area of minor priority as these countries have various other immediate problems to deal with and developmental needs to meet with. Many developing countries are of the view that taking up any obligation regarding carbon sequestration would reduce them to carbon colonies, and that they would be asked to compromise with their development to comply with the agreed terms of such projects. At the present stage of development, reducing energy consumption would mean compromising with the developmental potential of the country. What can be done at this stage is to ensure that the increased CO_2 emission due to continuous use of energy sources is mitigated continuously. This can be achieved by afforestation and reforestation activities. So the implementation of carbon mitigation projects must be integrated within the general national economic policies. The Kyoto Protocol also includes three market-based instruments known as the Kyoto Mechanisms that allow countries to earn or buy credits outside their borders. They are Clean Development Mechanism (CDM), Joint Implementation (JI), and International Emissions Trading (IET). These mechanisms enable Parties to access cost-effective opportunities to reduce emissions or to remove carbon from the atmosphere in other countries. While the cost of limiting emissions varies considerably from region to region, the benefit for the atmosphere is the same, wherever the action is taken. The developing countries can participate in two of these mechanisms, directly in CDM activities and indirectly in IET through CDM activities.

The Clean Development Mechanism (CDM) is a way to earn credits by investing in emission reduction projects in developing countries. International Emissions Trading (IET) will permit developed countries that have taken on a Kyoto target to buy and sell credits among themselves.

Forests and agricultural soils can remove and store carbon dioxide from the atmosphere. These carbon sequestration activities are called 'sinks'. Sinks can be enhanced through sustainable management practices in forestry and on farms. The Kyoto Protocol provides for inclusion of sinks as part of a country's strategy to meet its obligations. India signed and ratified the Protocol in August 2002. Since India is exempted from the framework of the Treaty, it is expected to gain from the Protocol in terms of transfer of technology and related foreign investments.

The best possible forestry scenario among the various options available needs to be identified such that the demand of wood and wood products is met optimally, besides mitigating carbon. After the energy sector, the focus of the Climate Convention and scientific literature is largely on the forestry sector for undertaking various projects under Land-Use, Land-Use Change and Forestry (LULUCF). Absorbing CO_2 from air and injecting it into the biomass is the only practical way of removing large volumes of GHGs from the atmosphere.

FORESTRY OPTIONS FOR MITIGATING CLIMATE CHANGE

In India, the current and projected rates of afforestation and reforestation under the baseline scenario are inadequate to meet the biomass demand projected for 2015.

So India will be required to select from two options. The first option is to reduce the biomass demand. This can be achieved by substituting the traditional methods by new innovative, technology driven energy options by investing heavily in building totally new infrastructure and ensuring that it percolates down to villages as well. This option would be much more feasible in urban parts of the country where direct dependence on fuelwood and biomass is less compared to rural India.

Second option available is to increase the biomass supply so as to fill in the currently prevailing demand-supply gap. This will require growing trees as crops very much like other commercial crops. As we would see further in the policy study, the government support would be marginal.

The information needed for assessing the potential of degraded and fallow lands, issues related to them such as extent of degradation, their ownership and possible future use has not been systematically calculated for the entire country.

Government, in order to promote forestry activities, especially commercial, will need to develop appropriate policies to promote participation of industry, farmers, commercial banks and external agencies.

DEMAND AND SUPPLY OF WOOD IN INDIA

INDUSTRIAL WOOD

Industrial wood includes all types of wood other than fuelwood. Various industrial sectors, such as pulp and paper, plywood, match industry, sports goods industry, railways, packaging industry, vehicle body builders, agricultural equipment as well as furniture and housing sector make up the demand for industrial wood.

The demand for industrial wood in 1987 was estimated to be around 27 million m^3 of which 19.61 million m^3 was from the industrial sector and 7.93 million m^3 from the household sector.

The projected demand for total industrial wood for 1996 was 64.4 million m^3, 73 million m^3 for 2001 and 81.8 million $m^3$for 2006. The demand for industrial wood is met through supplies from government forests and non-forest sources, such as farmlands and homestead gardens. Table below shows the domestic timber production trend.

Table : Wood Production Trend in India 1980-94 (figures in million m^3)

Production of	*1980*	*1983*	*1990*	*1994*
Round Wood (Total)	212.1	237.7	273.7	294.0
Fuelwood	192.4	215.6	249.3	269.2
Industrial Wood	19.7	22.1	24.4	24.8
Share of Saw/Ply Logs	15.2	16.7	18.4	18.4
Share of Pulpwood/Particles	1.2	1.2	1.2	1.2

Source: FAO (1996).

Other estimates suggest an annual removal of fuelwood alone to be more than 400 million m^3. However, the Annual Allowable Cut (AAC) from government forests is only 66.7 million m^3. The annual sustainable cut from private sources is estimated at 60 million m^3. Therefore the total legitimate availability is just about 127 million m^3. The shortfall is understood to be made good by over-exploitation of forest resources. Table below shows a more conservative estimate of domestic timber production.

Table : Wood Production Trend 1998-2000 (figures for fuelwood in million tonnes (Mt), rest in million m³)

Production of	*1998-99*	*1999-2000*
Fuelwood	4.1	4.0
Timber	2.0	8.0×10^{-1}
Poles	6.5×10^{-3}	92×10^{-3}
Pulp & Match Wood	50.4×10^{-2}	46.6×10^{-2}

Source: Forestry Statistics India

The liberalized Indian economy has posted healthy growth figures since the early 90s. This has led to multiplication of the demand for timber. Massive real estate development, increasing urbanization and growing middle and upper income classes are the major drivers of this demand.

On the supply side, however, the Indian timber market continues to be disorganized and dispersed. Serious restrictions on harvest of notified forest stands, and a weak private plantation sector have created a dependence on expensive imports. This has led to an eventual shift towards plastics, aluminium and steel in construction and furniture industries. The substitution is estimated to be upto 25% of the timber market. Traders of timber and manufacturers and traders of value-added products, already facing acute shortages of domestically harvested raw material, are further hamstrung by the supply of reliable and timely timber trade information.

Pulpwood is an important raw material used by the paper industry. It has been calculated that the country's pulp and paper industry obtained 41.7% of fibrous raw material from government-owned natural forests, 25.2% from farm forestry, 20.6% from the open market, and 12.5% from captive plantations.

Fuelwood

India is a major producer and consumer of fuelwood. In 1996, fuelwood consumption in India was 201 million tonnes (Mt), out of which 162 Mt (80.6%) was consumed in the household sector. The rest was consumed by commercial units (bakeries, restaurants) and industrial units (tobacco, gum, brick, ceramics). Almost 80% of rural and 50% of urban households are estimated to use wood for household energy supply. According to the National Forestry Action Programme, consumption of fuelwood increased by 0.5% year on year, over a sustained period. Felling of trees and lopping of twigs and branches in the natural forests have been the major sources of fuelwood. In addition, shrubs on degraded lands and roadsides, planting through social and farm forestry programmes and homestead gardens provided other sources of fuelwood in the country. In the year 1991, these sources contributed 70.5 Mt, 46 Mt, 40 Mt and 16 Mt respectively, adding up to a total consumption of 172.5 Mt.

The demand for fuelwood has continuously increased with the increase in population. While the use of fuelwood declined with the growth of the size of cities, in towns of less than 20,000 inhabitants, fuelwood accounted for 80%

of the total energy consumption. The rural population depends on fuelwood in addition to other biofuels, such as dung cake and crop wastes. Fuelwood is used as a source of energy in 71.7% households in the rural area and 32.7% in urban area. The average annual household consumption of fuelwood in the forested rural areas has been almost four times higher than the non-forest rural areas of India.

Demand-Supply Gap

The demand-supply gap in the country is 37-38 million m^3 in the case of industrial wood and a staggering 121 Mt in the case of fuelwood. A part of this gap is being met through unsustainable removals from government forests and other lands. Demand projections for the near future suggest that the situation may indeed deteriorate. Tables show the projected demand situation till 2011. The projection was made using the projected population for 2011 and the 1981 population to demand ratio.

Table : Use of Fuelwood, crop residues and dung (in 10^6 tonnes).

Year	*Fuelwood (Air dry)*	*Crop residues (Air dry)*	*Dung (fresh – about 87% moisture)*
1953-54	86.3	26.4	46.4
1960-61	99.6	30.6	54.6
1965-66	109.3	33.6	59.9
1970-71	117.9	36.3	64.6
1975-76	133.1	41.0	73.0
1986	157.0		
2011*	191.0		

* Projection

Source: Ravindranath *et al.*

Table : Timber Use and Projected Demand for India (in 10^6 m^3)

Type of Timber Demand	*Estimate (1986)*	*Projection (2011)*
Long-term use	13.76	25.85
Short-term use		
- Paper	6.57	
- Match industry	0.44	
- Packaging	6.81	
Total short-term use	13.82	25.96
Total timber	27.58	51.81

Source: Ravindranath *et al.*

WOOD IMPORT IN INDIA

India's forest products import bill, of which a major part is attributable to timber, is upwards of Rs. 5,700 crores. India is one of the largest importers of logs in the global trade. India and Japan accounted for 68% of the reported

International Tropical Timber Organisation (ITTO) log export volume in 2003. In fact, India is now the second largest importer of tropical log, overtaking Japan for the first time in 2003 with imports of just over 2.4 million m^3, up 10% from 2002 levels. In recent years, near stable prices for tropical log products were due mainly to the growing demand in India and China. Table below lists the major timber/value-added items imported.

Table : India's Major Wood Imports

ITC (HS) Code	*Product Description*	*Value (in Rs. Lakhs)*
1999-2000		
44011001	Fuel wood in Logs	13.90
44011009	Other Fuel Wood	6.46
44012100	Coniferous Wood in Chips or Particles	4.76
44012200	Non-coniferous Wood in Chips of Particles	6.14
44031000	Wood Treated with Paint/others	1111.05
44032001	Sawlogs and Veneer logs in Rough	38.92
44034901	Teak Wood in Rough	40290.70
44034909	Others	11799.89
44039100	Oak Wood in Rough	110.30
44039200	Beech Wood in Rough	244.25
44039901	Andaman Padauk (*Pterocarpus dalbergioide*)	3347.19
44039902	Bonsom (*Phoebe goalparensis*)	406.02
44039903	Gurgan (*Dipterocarpus alatus*)	5609.34
44039912	Sal (*Shorea robusta*)	29.20
44039913	Sandal Wood (*Santalum album*)	10.06
44039929	Other	122953.67
44069000	Other Wooden Sleeper	208.16
44071001	Douglas Fir (*Psudotsuga menziesie*)	26.47
44071002	Pine (*Pinus spp*)	21.75
44071009	Other Coniferous Wood Articles	163.06
44079200	Sawn/Chipped Wood of Beech	170.51
44079901	Sawn/Chipped Wood of Birch (*Betula spp*)	128.03
44079902	Sawn/Chipped Wood of Willow	295.73
44079909	Others	262.80
TOTAL*	5,77,354	

* Includes substantial import of plywood, building materials, cork, wood pulp, paper and printed material based on tropical/coniferous wood

Source: Forestry Statistics India.

Imports are mostly from Malaysia and Myanmar but with an increasing component from Africa. A 23% drop to 1.9 million m^3 in India's 2004 imports was mainly due to a decline in reported trade with Indonesia, because of Indonesia's restrictions on exports. The shift toward African timber is also partly because of growing restrictions on Asian log exports. Also, the dependence on tropical timber is changing, with not only India but also

Malaysia and the Philippines now sourcing substantial quantities of timber imports from non-tropical areas. Table below lists the major suppliers of timber to India.

Table : India's Major Timber Suppliers

Sl. No.	*Exporter Country*	*Quantity*
1.	Malaysia	15,49,357
2.	Myanmar	3,40,797
3.	Nigeria	99,677
4.	Cote d'Ivoire	95,939
5.	Gabon	65,110
6.	Guyana	26,369
7.	Indonesia	7,309
	TOTAL	24,48,365*

* Includes minor imports from non-regular suppliers

Source: ITTO.

India's tropical plywood production, based largely on imported tropical logs, is also rising rapidly. India's production soared 23% in 2002 to 1.6 million m^3 overtaking Brazil and Japan. It surged a further 10% to almost 1.8 million m^3 in 2003 and remained at this level in 2004. India is now the fourth largest tropical plywood producer in the world. Also, India, Indonesia and Thailand are the only significant wood pulp producing countries in the ITTO club. India produced 1.7 million $m^3$of wood pulp in 2003. Whereas Indonesia's pulp industry is based on fast-growing plantations as well as tropical hardwoods from natural forests and pulp imports, Indian paper and pulp makers have to depend largely on timber imports.

PRODUCTION FORESTRY IN INDIA

Definitions

In the context of the current study, production forestry may be understood to be the raising of block plantations on private lands, leased forest lands or leased community lands with the following purposes:

a) supply of timber as raw material to wood-based industry, and
b) supply of fuelwood/small timber for mitigation of the huge demand-supply gap in energy needs in rural areas.

The size or the ownership pattern of the land in question may vary. It may be useful to consider the following additional definitions (and explanatory notes) by Food and Agriculture Organization (FAO), Rome for a better understanding of production forestry.

Forest

Land spanning more than 0.5 hectares (ha) with trees higher than 5 metres (m) and a canopy cover of more than 10 percent, or trees able to reach these thresholds *in situ*. It does not include land that is predominantly under agricultural or urban land use.

Explanatory notes:

a) Forest is determined both by the presence of trees and the absence of other predominant land uses. The trees should be able to reach a minimum height of 5 m *in situ*. Areas under reforestation that have not yet reached but are expected to reach a canopy cover of 10 percent and a tree height of 5 m are included, as are temporarily unstocked areas, resulting from human intervention or natural causes, which are expected to regenerate.
b) Includes areas with bamboo and palms provided that the criteria of height and canopy cover are met.
c) Includes forest roads, firebreaks and other small open areas; forest in national parks, nature reserves and other protected areas such as those of specific scientific, historical, cultural or spiritual interest.
d) Includes windbreaks, shelterbelts and corridors of trees with an area of more than 0.5 ha and width of more than 20 m.
e) Includes plantations primarily used for forestry or protection purposes, such as rubber-wood plantations and cork oak stands.
f) Excludes tree stands in agricultural production systems, for example in fruit plantations and agroforestry systems. The term also excludes trees in urban parks and gardens.

Other Wooded Land

Land not classified as Forest, spanning more than 0.5 ha; with trees higher than 5 m and a canopy cover of 5-10 percent, or trees able to reach these thresholds *in situ*; or with a combined cover of shrubs, bushes and trees above 10 percent. It does not include land that is predominantly under agricultural or urban land use.

Production

Forest/Other wooded land designated for production and extraction of forest goods, including both wood and non-wood forest products.

Production Forest

Forest actually designated for production of forest goods *i.e.* where the extraction of forest products, usually wood and fibre, are the predominant management objective. It includes both wood and non wood forest products.

Productive Plantation

Forest/Other wooded land of introduced species and in some cases native species, established through planting or seeding mainly for production of wood or non wood goods.

Explanatory notes:

a) Includes all stands of introduced species established for production of wood or non-wood goods.

b) May include areas of native species characterized by few species, straight tree lines and/or even-aged stands.

The second set of definitions are borrowed from common technical parlance, with special reference to what is in vogue in entities in the wood-based industry, such as paper mills.

Farm Forestry

The practice of raising trees as crop, to primarily harvest and sell the timber without reference to the size of the holding(s).

Agroforestry

The practice of growing timber along with pure agricultural food and cash crop species and/or horticultural species

Finally, terms such as industrial plantations (with connotations similar to the first definition in this section) and tree farming are also to be found in the literature in this domain.

PLANTATION RESOURCES IN INDIA

Plantation resources have been generated and maintained in India over the ages. Emperor Ashoka (273-232 BC) is known to have had trees planted along long stretches of main roads in his vast empire. Emperor Shivaji (1630-1680) is reported to have encouraged plantation forestry within his empire. Sher Khan (1472-1545), who asserted his independence from the Mughal Emperor Humayun and built and ruled over a large empire, is known to have formally converted an old imperial highway spanning almost the entire north of the Indian sub-continent into the Grand Trunk Road and had large stretches of the roadsides planted with trees.

The earliest plantation of the colonial era in India is reported to be of a native species, teak (*Tectona grandis*), planted in 1840 in Nilambur, Kerala. Regular planting, mainly of teak, began in 1865 in many of the teak-growing central and southern provinces.

In 1910, *Eucalyptus* spp. was introduced in the Nilgiri Hills of the present Tamil Nadu. Planting of other native species was accelerated after the *taungya* system was introduced in 1911. These plantations, however, did not cover an extensive area until 1950.

Planned afforestation for soil conservation, industrial wood, fuelwood and fodder started in the late 1950s. The total plantation area to the end of 1972 was about 2.1 Mha. Establishment of plantations remained confined mostly to forest reserves until 1979. The plantation boom occurred when the social forestry projects (SFM Programmes) were launched in many states along with several other afforestation projects carried out with the assistance of external donors. The annual planting rate increased to about 1 Mha during 1980-1985. Most plantations have since then been established outside forest reserves in wastelands owned by the government or on community or private farmers' land. Plantation forestry received further impetus when a National Wasteland Development Board was created in 1985. The annual rate of planting increased to 1.8 Mha during 1985-1990. The area of plantations established during 1980-1990 was estimated by converting seedlings planted/distributed by a notional equivalence of 2,000 seedlings to one hectare.

Records of plantations established since 1991 are maintained for planted area and distributed seedlings separately, by the National Afforestation and Eco-development Board (NAEB) created in 1992 at the Union Ministry of Environment and Forests. The annual rate of planting since 1990 has been ranging between 1.4 to 1.6 Mha.

India has sizeable plantation areas of non-forest species. The total area to 1997 was about 15.3 Mha, of which rubber (*Hevea brasiliensis*) occupied $1.2x10^{-2}$ Mha and bamboo and cashew (*Anacardium* spp.) occupied 0.4 Mha and 0.1 Mha respectively.

Species Composition

A large variety of species are planted in the varied agro-climatic zones. However, detailed information about the composition of species in plantations is lacking. *Acacia* spp., *Eucalyptus* spp. and *Tectona grandis* occupy the greatest areas in the plantations.

Eucalyptus globulus, E. grandis and *E. tereticornis* are most common species, while among the Acacias, *Acacia auriculiformis, A. catechu, A. mearnsii, A. nilotica* and *A. tortalis* are common. Other commonly planted broadleaves are *Albizia* spp., *Azadirachta indica, Casuarina equisetifolia, Dalbergia sissoo, Gmelina arborea, Populus* spp. *Prosopis* spp., *Shorea robusta* and *Terminalia* spp. Survival rates of species differ and sal (*Shorea robusta*) is known for seedling dying-back disease.

Among conifers, *Cedrus deodara* and *Pinus roxburghii* occupy a major area. *Pinus patula* and *P. caribaea* have been planted to a limited extent.

Growth and Yield

Wood production from forest plantations at the national or sub-national level is not available. It is reported that productivity from plantations in general is quite low. For example, mean annual increment (MAI) for teak at the average rotation age of 58 years varies between 0.6 to 7 m^3/ha/yr with a mean of 2.5 m^3/ha/yr in Kerala, one of the major teak producing states. This recorded productivity may differ from real productivity ~~on account of not taking into account the~~ since smalltimber and fuelwood removal from forests is not taken into account. Productivity levels of some plantations, mainly of eucalyptus and poplar, raised by farmers under private ownership, is better.

Yield of selected species is as below:

Dalbergia sissoo

Rotation (years): 30 to 40

Mean annual increment m^3/ha/yr: 4 to 6

Eucalyptus spp.

Rotation (years): 10 to 20

Mean annual increment m^3/ha/yr: 8 to 12

Gmelina arborea

Rotation (years): 30 to 40

Mean annual increment m^3/ha/yr: 10 to 15

Acacia nilotica

Rotation (years): 20 to 25

Mean annual increment m^3/ha/yr: 3 to 4

Populus spp.

Rotation (years): 8 to 10

Mean annual increment m^3/ha/yr: 20 to 25

LAND USE AND AVAILABILITY

India's biomass demands may be met through afforestation and reforestation coupled with sustainable plantation forestry management practices. Large potential exists for afforestation in private lands, such as farm fallow lands and marginal croplands. This chapter contains information on forestland, wastelands and land use patterns.

Present Outlook

In the absence of a robust industrial plantation movement, several types of relations between woodbased industry and the farmer community are seen to have evolved:

- Supply of free or subsidized seedlings
- Bank loan schemes
- Leasing or share cropping schemes
- R&D and commercial sale of improved clonal planting stock Saigal, S., Arora, H., Rizvi, S.S. (2002). The new foresters: The role of private enterprise in the Indian forestry Sector. Ecotech Services, New Delhi and International Institute of Environment and Development, London.

In the early 1990s, many private companies started teak plantations by raising funds from public sources. They promised very attractive rate of return. Their calculation was based on the assumption of high volume per tree on a 20-year rotation, which was found unrealistic and wrong. The plantation sites were mostly degraded and less productive and the planting material was also not of genetically superior quality. On scrutiny of accounts of some planting companies by the Securities and Exchange Board of India (SEBI), it was found that the companies depended on the fresh investors and not on interim returns of teak plantations to honor the commitment of earlier investors. The companies also could not prove their claims of high productivity, as a result most of the companies were closed down during late 1998. The total area brought under teak plantation by them was around 5,000 ha.

FACTORS AFFECTING GROWTH OF PRODUCTION FORESTRY

Several factors constrain the growth of production forestry in India:

- High cost of credit and lack of access to credit
- Long gestation period, particularly for long-rotation forestry
- Absence of market institutions
- Regulations on planting, harvesting, and transportation of trees
- Import of timber, pulp, and paper
- Lack of high yielding genetic planting material and silviculture practices
- Absence of forestry extension services to farmers
- External funding largely to forest department, not for farmers or plantation companies
- No foreign direct investment in forestry.

The disaggregation of production forestry sector is possibly also reflected in surveys and assessment figures thereof. Only the annual statistics on plantations done under different schemes and the cumulative plantation areas since 1951 are available. Monitoring and inventory of plantations is inadequate and as a result the actual area of existing plantations is uncertain.

Plantations established after 1985 constitute more than 70 percent of the total. Part of the annual plantation target (35 to 40 percent), particularly after 1985, is achieved by the distribution of seedlings.

Since 1992, NAEB had started survival assessments of first-year plantations in limited areas (about 10 %), but the results of the assessment are not used to correct the area figures reported.

PRODUCTION FORESTRY AND CARBON SEQUESTRATION

RATIONALE FOR CARBON SEQUESTRATION

Climate change (generally alluding to global warming) refers to the variation in the climate of the Earth due to various factors such as variation in solar radiation, the Earth's orbit and greenhouse gas concentrations. Of these factors, increasing concentrations of greenhouse gases (predominantly carbon dioxide, CO_2) is attributed to anthropogenic activities. The carbon reservoir (carbon dioxide sink or CO_2 sink) is increasing in size, and the global carbon stock is projected to increase by a net 290 Gt (Gt = thousand million tonnes) due to CO_2 increases, climate change and vegetation redistribution.

Carbon sequestration describes the processes that remove the carbon from the biosphere. The IPCC concluded that the 'cumulative amount of carbon that could potentially be conserved and sequestered over the period 1995-2050 by slowing deforestation (138 million ha) and promoting natural forest regeneration in the tropics (217 million ha), combined with the implementation of a global forestation programme (345 million ha of plantations and agroforests) would be about 60-87 GtC-, equivalent to 12-15 percent of the projected cumulative fossil fuel and deforestation emissions over the same period'. Many other ways have been suggested to mitigate the increase of carbon in the atmosphere. Land use, Land-use Change and Forestry (LULUCF) and agriculture sequester carbon in forests and agricultural lands. The other option is to decrease emissions of CO_2 and other Greenhouse Gases (GHGs) by substituting fossil fuels with biomass substitutes. While this would be a better way to control CO_2 emissions, forestation is a tool for bringing about a net reduction in atmospheric carbon levels.

CURRENT STATUS OF CARBON SEQUESTERED IN INDIAN FORESTS

The forestry sector has the potential to sequester carbon in addition to sustain its carbon. The total carbon pool of forests ~~are~~ is assessed from estimates of biomass from growing stock of forest stratum. IPCC IPCC (1995). Guidelines for greenhouse gas inventory workbook, Volume 2, Module 5 - Land Use Change and Forestry, Report prepared by UNEP, OECD, IEA and IPCC, 5.1-5.45.

Prescribes a default value of 0.95 as the conversion and expansion factor to convert volume (in cubic metres) to biomass (expressed as tonnes of dry matter). IPCC also prescribes a factor 0.45 for converting the biomass into carbon content. Other researchers have used similar values, *e.g.*, Chaturvedi assumed the factor to be 0.48 and Bhadwal and Singh assumed the factor to be 0.5.

The major grouping of Indian forests (based on climate) are:

1. tropical
2. montane subtropical
3. montane temperate
4. alpine.

These are subdivided into 16 sub-types based on Champion and Seth classification. Of these sub-types, majority of the area is under tropical moist deciduous and tropical dry deciduous forests. The total forest cover of India has been around 64 Mha since the 1980s. Dhadwal and Nayak (As cited in Lal and Singh) estimated the total carbon pool of Indian forests to be 1994 million tonnes (Mt) for the year 1985. Ravindranath and Somashekhar estimated the total carbon pool for Indian forests to be 4179 Mt for the year 1986. In a more recent research study, Lal and Singh estimated the total carbon pool for Indian forests for the year 1995 to be 2027 Mt. Dense natural forests are better stores of carbon than open natural forests. A study of natural forests of Madhya Pradesh by Pande estimated the total carbon pool in standing crop as 363 Mt for dense forests and 80 Mt for open forests. Open forests, therefore, have a potential to sequester more carbon.

It is also to be noted that the annual productivity of Indian forests have been reported to increase from 0.7 m^3 /ha in 1985 to 1.37 m^3 /ha in 1995.

Estimates of Future Carbon Sequestration

Prasad *et al.* have modeled the land use changes and forestry data of India from 1997 to 1999 and suggested that Indian forests would be a potential sink for 0.94 Gt of carbon over a period of time, with an increase in dense forest area of about 76 Mha and decrease of about 3 Mha and 5 Mha in open and scrub forests, if similar land use changes that occurred during 1997-1999 would continue. The study bases its observations on final outcomes on the land use patterns (and changes therein) observed between 1997 and 1999. The authors use stochastic modeling techniques to determine relationships between land usage and magnitude of carbon stored in forests and soils. They extend the model to map net carbon flux between the land and the atmosphere. An entire series of studies on methods for examining the future consequences of changes in land usage and carbon flux patterns are also referred to and evaluated as sources of suggested carbon mitigation strategy options.

A study by Bhadwal and Singh uses the Land Use and Carbon Sequestration (LUCS) model with software developed at the World Resources Institute, Washington, DC, USA. Indian agriculture and forest statistics were systematically integrated with geographic and demographic data. Three distinct scenarios – business as usual, conservation and plantation forestry were considered. It was estimated that under plantation forestry, 7 billion tonnes of carbon will be sequestered during 2000-2050.

In an earlier study by Ravindranath *et al.*, a sustainable forestry scenario and a commercial forestry scenario were developed with the aim of meeting the biomass demands through plantation forestry and conserving the forests. Comparison of the commercial forestry scenario over the baseline scenario (in the period 2000-2012) estimated that an additional carbon stock of 78 x 10^6 Mg C would be sequestered apart from meeting all the incremental biomass demands (estimated for 2000-2015).

Economics of Production Forestry

In the study mentioned above, Ravindranath *et al.* considered the cost-effectiveness of four potential forestry activities to meet the biomass demand of the country. These activities were:

1) short-rotation forestry (afforestation to meet the fuelwood and industrial wood requirements),

2) long-rotation forestry (afforestation to meet the structural (sawnwood) requirements),
3) forest regeneration (reforestation to regenerate and reclaim the degraded forest lands through protection and promotion of forest succession), and
4) forest protection (to conserve the biomass of natural forest by halting deforestation).

The mitigation potential on a per-hectare basis for the period 2000-2030 was lowest for short-rotation forestry (at 25 Mg C/ha) and highest for forest protection (at 176 Mg C/ha). The life cycle cost of mitigation was lowest for forest protection (USD 0.39 /Mg C) followed by forest regeneration (USD 0.99 /Mg C), long-rotation forestry (USD 11.69 /Mg C), and short-rotation forestry (USD 18.8 /Mg C).

Status of Production Forestry in India

The area under plantation in India has been growing over the past decades. The rate of afforestation in India was 1.67 Mha/yr during the 1980s and 2 Mha/yr in the 1990s. Agroforestry also has the potential to store carbon and remove CO_2 through enhanced growth of trees and shrubs and has been demonstrated as a promising mechanism of carbon sequestration in India. Carbon sequestration in Indian agroforests varies from 20 tC /ha/yr in Uttar Pradesh to a carbon pool of 23-47 tC/ha/yr in tree-bearing arid agroecosystems of Rajasthan. The average sequestration potential of agroforestry has been estimated to be 25 tC/ha/yr over 96 million ha of land in the country.

With the aim to sequester carbon rapidly, it is best to grow fast-growing species with shorter rotation periods. In India, the choice of species for plantations is dictated by market demand for eucalyptus and poplar poles. Other fast growing species are *Albizia, Cassia siamea, Leucaena leucocephala, Casuarina equistifolia*, and *Dalbergia sissoo*.

Afforestation in India is generally carried out under the social forestry programme. This includes growing trees on community commons, public or government owned land and aveanue plantations. Farm forestry that includes raising plantations of eucalyptus, casuarina, teak and other commercial species on farmland, or on bunds is considered agro-forestry. Eucalyptus species and *Acacia auriculiformis* are grown under the reforestation programmes that included the following two categories of plantations:

1. plantations on village commons and degraded forest land that were carried out to meet the biomass need, such as for fuel and small timber.
2. farm forestry on farmers' own land to meet industrial demand

In India, plantations can also be categorized as:

1. short rotation plantation forestry (species such as *Eucalyptus, Acacia auriculiformis* and *Casuarina equisetifolia*), and
2. hardwood plantations (species such as *Tectona* and *Pinus*).

Broadly speaking, the major plantation species are poplar and eucalyptus in north India, subabool in central India, and eucalyptus, *Acacia auriculoformis* and rubberwood in south India. The productivity of forest plantations under the national afforestation programme is significantly higher compared to natural forests. A national level study of forest plantations in India estimated the mean anuual productivity (of woody biomass) to be 3.2 tonnes per hectare per year (t/ha/yr). There are a few location-specific studies on productivity of forest plantations and the range of productivities recorded vary from 1.2 to 8.2 t/ha/yr, with the national mean productivity of about 3.1 t/ha/yr. This variation is not surprising since the case studies were location-specific and ranged from farm forestry in semi-arid regions to plantations on degraded forest lands in heavy rainfall zone. It productivities could range from 2-5 t/ha/yr in semi-arid regions and 5-10 t/ha/yr in sub-humid regions. One should also note that the productivity of any forest plantation would depend on the age of the trees.

While the species choice and productivity is largely determined by factors such as climate and soil, the productivity can be potentially increased by improvement in genetic stock, application of fertilizers and irrigation. A preliminary estimate of potential productivities by Ravindranath and Hall places it in a range of 20-35 t/ha/yr. The total carbon flux of Indian forests can be calculated based on the total carbon uptake and emissions into the atmosphere. For the year 1986, Ravindranath *et al.* estimated the total carbon emissions from forests to be 64 Mt and the total carbon uptake to be 69 Mt. They have estimated that carbon uptake from the area brought under tree plantation and the existing forest under forest succession will offset the gross carbon emission in the country, and projections for the year 2011 show large potential for net sequestration of carbon.

POLICIES IN INDIA ON PRODUCTION FORESTRY

NATIONAL AGRICULTURAL POLICY

National Commission on Agriculture (NCA) recommended a change over from the conservation-oriented forestry to more dynamic programme of production forestry. According to the Commission, production of industrial wood would have to be the main reason for the existence of forests and should be project-oriented and economically feasible.

The NCA recommended against leasing forestlands to industry and instead wanted government to allocate them forest raw material through enhancing commercial forestry. For the purpose of commercial forestry, it further recommended that out of 64 million hectare (Mha) of forest lands available in India at that time 48 Mha should be brought under production forestry and remaining 16 Mha for biological diversity.

NATIONAL FOREST POLICY (NFP)

NFP, 1952 acknowledged the need for sustained supply of timber and other forest produce to meet the developmental need of various industries. The policymakers had two major objectives in mind while framing NFP 1952. First to ensure that forests were preserved and managed on a sustainable basis, and second to use them for meeting national interest as a source of timber. Since both these objectives somewhat opposed each other, NFP came up with a concept of '*treelands*'. It identified the scope for involving state governments as well as various other institutions, such as, Defence, Rail-ways, Public Works Departments, Universities and Colleges, Boards, Municipalities and other local authorities, associations and institutions in the process by converting the land at their disposal into treelands. To ensure that the needs of forest-based industries are met without putting undue pressure on forests the need to substitute tree-species of commercial importance in place of inferior tree–species was advocated. The policy also mentioned involving both the industries and individuals in a bigger manner and emphasized on considering commercial and industrial interests for establishing closer contacts and bonding between Forest Research Institutes and industries utilizing timber and forest products.

Forest Conservation Act, 1980

The Act put a ban on the de-reservation of forests or use of forestland for non-forest purpose. The Act of 1980, consequently, brought the subject of forests from state list to concurrent list enabling the parliament of India to look into the matter of forests. After the Act, no state government could use forestland for non-forest purposes without prior endorsement by the central government. This put up a check not only on the conversion of forests but also brought to halt the fragmentation of the remaining forests. This ensured that a local party could not take the decision of diverting the forests for taking up any political mileage. The centre acted as an independent agency, to which the state governments and their decisions were answerable.

National Forest Policy, 1988

The NFP, 1988 looked upon forests not as a source of raw material for commercial purposes, but primarily for conserving soil, water and biodiversity besides meeting subsistence requirements of the local people. The policy clearly mentioned that the economic utilities coming from the forests would be secondary to this prime aim.

The NFP 1988 shifted the official focus from fuelwood and timber to the management of forests primarily for their services. It made it imperative to ensure that the forests are not only conserved but also their cover and productivity is increased so as to meet the demand of goods and services coming from forests. The policy, while addressing the production forestry programme, expressed its concern in plugging the increasing demand-supply gap of fuelwood and meeting the national needs. But it explicitly mentioned that no such activity should result in any sort of clear felling of already existing natural forests. It expressed the desire that such programmes should help the country meet with its objective of achieving one-third forest cover.

The policy put a thrust on meeting two important concerns of utilizing the wastelands and increasing tree cover by promoting major forestry programmes. To reduce the pressure on forests, it also aimed at promoting substitution of wood besides encouraging better and efficient utilization of forest produce. Since land plays an important role in taking up plantations, it advocated that a slight modification in land laws should be taken up so that the process becomes smooth for the investors. At the same time, it maintained that any sort of leasing should be in coherence with the existing land ceiling act. The policy discouraged the earlier trend of meeting raw material requirements of forest-based industry by operating on natural forests. It directed these industries to meet their requirements by establishing a direct relationship with individuals who had the capacity to do so. The industry was encouraged to chip in by contributing its resources and expertise at various levels during the total project duration. The policy also gives more importance to native species over the exotic species. It discourages the use of exotic species without prior scientific trials to establish that there are no adverse effect on the environment.

National Forestry Action Programme, India (NFAP), 1999

The NFAP aimed to prepare an action plan for next 20 years, in conformity with NFP 1988. It is a comprehensive strategic plan to address the issue underlying the major problems of the forestry sector and to reverse the process of degradation for sustainable development of forests. The consumption of fuel-wood in India was reported to be about five times higher than what could be sustainably removed from forests. Further, a large tract of forestland was claimed (nearly 4.3 Mha) for undertaking industrialization and developing infrastructure necessary for meeting the development goals of the nation. The NFAP aimed at meeting the requirements of forest based industries without compromising the conservation and protection of natural forests.

The basic purpose of NFAP was to establish direct linkage between the NFP and the National Five-Year Plans (FYP). In the past, a comprehensive and constant programme structure for forestry was found missing. Every plan had its own programme structure. So it was difficult to find linkages and establish trends. Although plans had specified objectives and programmes, the main activity under most of them was tree planting. NFAP recognized the capability of plantations to help conserve the natural forests by providing an alternative source for forest products. The Programme was also optimistic regarding the capability of plantations to earn foreign exchange, besides meeting the domestic requirements of the country. The Programme admitted that despite the policy addressing direct relationship between industry and farmers, the government did not adequately support private initiatives. It raised concern on the inability of the government to provide these initiatives with relevant research, extension, technological packages, input delivery, and market information or credit facilities. It felt and remarked that it was imperative to encourage small operators, keep them interested in sustainable forestry development and understand their needs adequately.

NFAP suggested looking at these plantations as a means of raw material for industries or for meeting energy requirements. It intended to conserve and rehabilitate 31 Mha of degraded forests (less than 40% crown density) in addition to bringing 29 Mha of non-forest land under plantations. It advocated the use of different strategies for both these targets. The problem of degraded forests was to be tackled by involving local communities, especially the areas near villages through Joint Forest Management (JFM). For non-forest land, market-oriented massive tree plantation drive involving multiple agencies was proposed. The NFAP came up with a strategic approach to meet its objectives. The approach had five interrelated 'strategic areas' which formed the very basis of the Programme. These approaches were:

- Protect existing forest resources
- Improve forest productivity
- Reduce total demand
- Strengthen policy and institutional framework
- Expand forest area

The state governments while preparing the State Forestry Action Programmes (SFAPs) incorporated all planned activities as part of these activities.The NFAP recommended that for the sustainability of forests, the productivity of forest plantations was to be increased at least 3 to 5 cubic meter per ha per year (m^3/ha/yr) by promoting regeneration and enrichment of plantations. Plantations to be carried out on all categories of wastelands were also suggested. Keeping in mind the fuelwood requirements of the country, emphasis was paid on taking up plantations of fuelwood species on non-forest wasteland. Strengthening the institutions for people's participation in protection and development of degraded and fringe forests was emphasized.

Regarding the financing part of the programme, the sources of funds were grouped under four broad categories, namely, domestic public financing, domestic private financing, external public financing, and external private financing.

NATIONAL ENVIRONMENT POLICY

National Environment Policy (NEP), 2004

NEP emphasized the need to develop a strategy to meet the goal of raising the forest cover of the nation to 33% by 2012. It promoted the involvement of non-forestry sector as well. For increasing the forest and tree cover of the nation, multiple stakeholder partnerships have been recognized where each stakeholder would have his role clearly defined. For achieving the aforesaid target of 33% forest cover, the NEP 2004 suggested that the afforestation activities should be taken up on degraded land, wastelands as well as private land holdings. Key elements of the strategy would include: (i) implementation of multi-stakeholder partnerships involving the Forest Department, local communities, and investors, with clearly defined obligations and entitlements for each partner, following good governance principles, to derive environmental, livelihood, and financial benefits; and (ii) rationalization of restrictions on cultivation of forest species outside notified forests, to enable farmers to undertake social and farm forestry where their returns are more favourable than cropping. The policy called for development of a strategy to achieve the targets set for the eleventh five-year plan.

The problem of climate change and its underlying causes in global warming and increased usage of GHGs has been addressed in the Policy. It identifies deforestation in the country as one of the reasons for climate change expresses concern that India and other developing nations would be affected the most by climate change. The current level of emission in India is substantially lower than that of the developed nations; however, India's economic growth could result in an increase in GHG emission. However, government policies favouring renewable energy and afforestation projects as well as the growth of less energy intensive service sectors would result in a decrease in the level of emissions.

PLANNING COMMISSION REPORTS

Report on Leasing of Degraded Forest Lands, 1999

The Report was of the view that the degraded forestland should not be leased out to private entrepreneurs. In accordance with NFP 1988, it recommended that industries needing forest raw material should establish contact with farmers. The government would lease land to Forest Development Corporation (FDC) who, in turn, would enter into proper MoU with the user agency without leasing the land to them, as per the GOI guidelines of 1994. This MOU would give the user agency the right to undertake afforestation and take a fixed percentage of forest produce at the time of harvesting.

The Report recommended leasing barren land far away from habitations, available in plots of 1000 ha or more, and of no use to the villagers. These are desert or ravine or saline lands, which require huge investment before they can be made productive. The Report gives reference of Investment Promotion Scheme of Ministry of Rural Development, Government of India wherein such enterprises would be entitled to 25% subsidy on their capital investment. This means that the land needed for production purposes could be mobilized for meeting the demand of industry provided the industry is ready to put in huge investment besides following the guidelines. The report estimated that 33 Mha of degraded non-forest lands and 27 Mha of degraded forests (a total of 60 Mha) would be available for tree growing.

The Report also addresses the issue of difference between barren and degraded lands. Uncultivated public land was classified under these two categories. Barren land were defined as having forest/tree cover less than 10% whereas degraded land were classified as having forest/tree cover between 10 to 40%.

Greening India for Livelihood Security and Sustainable Development, 2001

Greening India Programme proposes to cover 43 Mha degraded land (15 Mha of degraded forestland, 10 Mha of degraded irrigated land, and 18 Mha of degraded rainfed land). It recognises that there exist numerous problems that make the implementation of agroforestry programmes difficult:

- Cumbersome legislation with respect to tree felling, wood transportation and processing.
- Lack of market information and infrastructure
- Dearth of appropriate agroforestry models
- Absence of economic security and incentives for tree growers
- Lack of extension training and demonstration
- Non-availability of quality planting material
- Unfavourable Import and Export policy

The Report highlighted the lessons learnt from the farm forestry programme:

- Tree planting should not be undertaken in uncultivable lands, but in cultivated-field and homesteads.
- All wastelands, non-forest areas and degraded forestlands should be brought under silvi-pastoral system with suitable species for fuel and fodder production. This can be addressed through JFM.
- Coastal land through afforestation, ravines and sand dune area for land reclamation, area under mining leases, water-logged areas can be targeted for converting into forested land.

Five-Year Plans

Forestry has also been covered from the first five-year plan, though it remains neglected most of the time in the planning process. Of the total fund allocated to various sectors, funds for forestry in India hovered around 1% only. The expenditure on forestry sector has increased substantially since sixth five-year plan in accordance with the increase in afforestation activities undertaken to meet the desired targets. The sixth five-year plan coincided with the FCA 1980. So the increase in spending on forestry sector can be attributed to a shift in governmental policies. In the fifth plan, the spending was a meagre Rs.107.28 crore, which increased to Rs.15964.06 crore at the end of ninth five-year plan. The area afforested under the same time period has increased from 12.21 hectares to 80.50 hectares. The comparison shows that the spending has increased at a much higher rate than the afforested area. The area afforested under seventh, eight and ninth year plan has been almost same, but the spending has almost tripled in same time period. This shows that growing trees alone will not make these afforestation activities successful. Proper care and maintenance of the area afforested needs higher allocation of money. In fact, money spent in ninth five-year plan was close to 90% of what had been spent till date in the country on these activities. This sends positive signals regarding finance to the project proponents and implementers about the success of afforestation.

Tenth Five-Year Plan (2002-2007)

The current five-year plan targets to increase the forest/tree cover to 25%. It agrees that no strategy would be successful unless and until the basic needs are met. Recognizing the role played by JFM in regeneration of degraded forests, the plan recommends taking further steps in identifying its strengths and weaknesses, so that the area under implementation can be increased. Raising concern on high imports of round timber and other forest produce, it expresses a desire for reversing this trend. This could be achieved by utilizing community land, degraded forests or private farmlands of the country. Implementing the same would mean removal of government subsidies, regulation of tariff on imports and other such policy modifications, which will make the plantation activities more desirable among farmers and village communities. Promotion of technology, credit support, developing marketing infrastructure and providing extension and training to interested farmers has also been recommended. Policy recommendations for tackling the problems related to constraint of felling, transportation and selling of forest produce which act as disincentive for growing trees are also suggested. This asks for creating a favourable policy environment, which promotes positive interaction among various stakeholders.

Keeping up with the challenges facing the society, plan proposes to utilize the wastelands and degraded lands for taking up CDM projects. This would help in utilizing the wastelands besides generating additional incentives for taking up plantations. Emphasis has been put on growing species such as *Jatropha curcas* and *Pongamia pinnata*, which also grow naturally and can be utilized for generating bio-diesel, although success of these schemes are yet to be fully realized.. These plantations are cost-effective and easily replicable. To take up the afforestation activities more seriously all programmes have been merged under a single scheme called 'National Afforestation Programme (NAP)'. It is being operated through Forest Development Agencies (FDAs). Similarly, another programme named National Action Programme to Combat

Desertification under UN Convention to Combat Desertification (UNCCD) has been taken up by MoEF. A 20 years' comprehensive NAP to combat desertification in the country was prepared with the following objectives:

- Community based approach to development,
- Activities to improve the quality of life of the local communities,
- Raising awareness,
- Drought management preparedness and mitigation,
- R&D initiatives and interventions which are locally suited,
- Strengthening self-–governance leading to empowerment of local communities.

For the Tenth Five-Year Plan, it has been proposed to initiate activities that include, among others, assessment and mapping of land degradation, drought monitoring and early warning system groups, drought preparedness contingency plans, and on-farm research activities for development of indigenous technology, etc.

The growing demand of raw material from our natural resources is threatening them. Envisaging the threat to the natural resources due to growing demand of raw material,is, the plan Plan proposes to take up plantation, which would reduce pressure on the natural forests and reverse the negative impact of deforestation while meeting the increasing demand. The Plan proposes to encourage agro-forestry by promoting technology, extension, and training, credit support, marketing infrastructure, etc., and providing a policy environment, which assures the farmers of a remunerative price. It also proposes to removes the constraints of felling, transport and marketing of forest produce from private holdings in different States and to formulate a common guideline for this purpose. The current mean annual increment (MAI) of forest plantations varies from about 2 m^3/ha/year for valuable timber species to about 5-8 m^3/ha/year for eucalyptus and other fast growing species. Generally, MAI is around 10 m^3/ha/year in good quality plantations in various countries. This poor performance of forest plantation remains a great concern in the area of policy decisions. The Plan suggests measures including appropriate site selection, site-species matching, planting of elite clones, proper maintenance and protection, timely tending, thinning, irrigation, application of manures and pesticides, etc. for improving the productivity of plantations. For improving the utilization of plantation, the plan identifies Indian Plywood Industries Research and Training Institute (IPIRTI) to be actively involved. IPIRTI can also help develop programmes that will not only fulfill the demand, but also provide with more wood substitutes.

POLICIES GOVERNING FOREIGN TRADE IN FORESTRY PRODUCE

Timber price in India has been increasing at the rate of 15% per year, making import of timber more attractive, and Tthe current import policy allows duty free import of timber and pulp. In order to help wood-based industries meet their requirements, the government removed the trade barriers and liberalized the import of timber on Open General License (OGL). For industries, it was a welcome step and had an additional benefit of growth of industries in the coastal belt. This policy change led to increased import of timber. However, the produce from the imported timber is sold mainly in domestic market and there is practically no export of products against import. Export is less than one tenth of import. India's foreign exchange reserve as well as interest of tree growers has been put under an unwanted pressure. It created a sense of insecurity among tree growers of the country who now had to be satisfied with a smaller share of the market, as global players offered attractive deals to capture a bigger share of the growing market. Despite tough competition from the forestry sector and importing agencies, farmers are still supplying 50 per cent of wood supplies from their holdings. The importance and potential of agroforestry and several other such models has not been realized.

The acceptance of timber from plantations as raw material in wood based industries has opened new avenues for these farmers. However, to give a boost to tree growing activity as a means for import substitution, certain policy initiatives were felt necessary and were recommended.

- To impose heavy import duty on such wood products, which are/can be produced within the country to meet domestic needs.
- Enhance R&D efforts to manufacture quality products from plantation grown wood and other renewable fibres for import substitution.
- Export of wood and natural fibre based products promoted through incentives and simplification of procedure.
- Evolution and implementation of minimum mandatory material and product standards.
- Mechanism for collection and dissemination of information regarding import, export, prices and trade of timber needs to be developed.

The import of various wood-based commodities has been quite high than the exports both in terms of quantity and money involved. Though the exports have increased a bit compared to import in year 1999-00 over 1998-99, huge gap still remains. Besides, the farmers would find it difficult to make a financial comeback to take up this activity again if their initial investment does not return enough profit. Moreover, import of wood reduces the option of generating additional employment opportunities in the country, which would be easily taken up by the low-income class. A general reason cited for this import has been to reduce pressure on forests.

Production forestry, farm forestry, agroforestry, if utilized to their hilt can do away with this import, which as of now puts a burden of Rs. 8,000 crores annually. Moreover, this would also help achieve the goal of one-third forest cover. Now the industries are also interested in taking up these activities by building relationship with farmers. The north-western part of the country took these efforts on a large scale, and massive plantations of eucalyptus and poplar were raised. However, a lack of organized market along with government policies, which make the process of harvesting trees and their transportation cumbersome, did not allow this effort to get the optimum revenues. Wood import under Open General License (OGL) made the proposition of importing wood raw material more attractive. Generally, this imported wood comes from natural forests outside the country. In many cases, the harvesting procedure is unsustainable. In a way, these steps save Indian forests at the cost of forests in other wood-exporting countries. In today's scenario of climate change, a phenomenon occurring at one place can have complex implications at far off distances as well. So the problem gets transferred instead of getting solved. So the solution lies in raising more trees within the country. This will provide incentive for local farmers and a disincentive for those countries indulging in these unsustainable activities by taking benefit of imperfect market conditions.

Financing Schemes

Institutional funding is very important in areas where a farmer-industry relationship is to be established. In order to promote the afforestation activities, the central government has been spending money through the five-year plans. Schemes such as Investment Promotional Scheme have been instrumental in initiating and promoting production forestry throughout the nation, though not as successfully as they were initially conceived. The experience in the last two decades show that institutional funding has been minimal in forestry programmes. These are some important reasons:

- Dearth of technical and economic data on different farm forestry models. This limits the ability of banks to evaluate bank ability of various farm forestry projects.
- Producers and banks often find the associated risk unacceptable. There are no insurance system to guard against the loss to producers and banks arising from various natural calamities.
- The lending banks do not have adequate capability to assist in formulation and appraisal of projects for farm forestry.

Investment Promotional Scheme

The Scheme was launched in the year 1994-95 in order to stimulate involvement of the corporate sector/financial institutions, etc. to pool in resources for development of non-forest wastelands. The principal objectives of the scheme were:

1. To facilitate/attract/channelize/mobilize resources from financial institutions, banks, corporate bodies including user industries and other entrepreneurs for development of wastelands in non-forest areas belonging to Central and State Governments, panchayats, village communities, private farmers, etc.

2. To promote group of farmers belonging to different categories, namely, large, small, marginal & SCs/STs for bringing wastelands under productive use.
3. To facilitate production and flow of additional biomass including farm-forestry products used as raw material inputs for different types of industries.
4. To facilitate employment generation through land development and other allied land based and related activities including plantations.

The Scheme was restructured to make it broad-based and circulated to all the states and other concerned in August 1998. Under this Scheme, Central Promotional Subsidy was limited to Rs. 25 lakhs or 25% of the project cost for on-farm development activities, whichever was less, subject to condition that the promoter's contribution in the project shall not be less than 25% of the project cost. The projects promoted by the Scheduled Commercial Banks (SCBs), Regional Rural Banks, Land Development Banks and Cooperative Banks were eligible for promotional grant/subsidy under the Scheme. Under the Scheme, 41 projects covering an area of 1435 ha with a total cost of Rs.16.88 crores (firmed up by the bank) and subsidy of Rs. 1.09 crores have been sanctioned up to March, 2004. Because of slow progress, the Scheme was discontinued in 2003-04.

Tax Deductions

Tax deduction is an important financial incentive for production forestry. The Government of India announced tax deduction to companies for carrying out projects of softwood plantation on degraded non-forest land.

LAND LEASE AND LAND CEILING ACTS

Land Ceiling Acts

In India, land falls under state list and is governed by the acts passed by states after independence. Accordingly there is no uniform binding on the land holdings by individuals in various part of the country. Records in the country indicate disparity and inequality in distribution of land. Large number of cultivators owning relatively less land, while big landowners, smaller in number owning larger acreage of land. It leads to disparities in the incomes in the rural areas. In view of this, our leaders in the earlier days thought of undertaking land reform measures. The ceiling on land holdings was intended to meet the land needs of the landless, reduce inequalities in land ownership so that it may lead to development of co-operative rural economy, and enlarge self-employment in owned land as distinguished from subletting and tenant cultivation.

The ceiling legislations were initiated in many parts of the country in the late 50s and early 60s. Jammu and Kashmir was the first state in the country to pass an act. West Bengal and then Himachal Pradesh followed suit.

Maharashtra passed this Act in 1961. However, the progress of ceiling legislation was disappointing till 1972. It was found that only about 23 lakh acres of land was declared surplus. Of this, about 13 lakh acres were redistributed. In Bihar, Karnataka, Orissa and Rajasthan, no land was declared surplus. It was mainly due to partitioning of land or *benami* transfers. Therefore, new guidelines were suggested and accordingly, 17 states amended the ceiling legislations. The range of ceiling varied from state to state.

Leasing of non-forest wastelands

Non-forest wastelands distant from villages usually lie barren, and are available in large chunks. In the past, some state governments have offered barren lands, such as desert lands of Rajasthan and saline lands of U.P. and Gujarat, on lease to industry, but the response has been lukewarm. The scheme by the Rajasthan government was initiated in 1990-91, wherein a minimum of 2000 acres was offered on a long-term lease, whereas the scheme in U.P. was initiated in 1979, but given up after a few years for want of proposals.

The Gujarat government since 1994 is offering land upto 2000 acres for reclamation, but there has been no positive response from forest based industry. Under this policy non-cultivable wasteland including sandy and saline land can be granted to NRI companies, co-operative societies, Indian industrial houses, public limited companies and other persons for developing the land or plantation.

So far only one industry has come forward and has been granted 316 acres for plantation. Similar facilities for granting leases to industry beyond ceiling limits exist in Tamil Nadu. Therefore, the complaint of industry that it is being denied the opportunity to rehabilitate wastelands or that land ceiling laws are a barrier to raising plantations is unfounded. States like Rajasthan, Gujarat, and Madhya Pradesh where most wastelands are located have adequate provisions in their laws to grant exemptions from ceiling in deserving cases. It has been recommended that private entrepreneurs need to consider reclaiming barren non-forest lands remote from habitations through modern technology.

One-time sale of quality seedlings by the industry to farmers has worked best in India. It is a simple and quick process with no risk or need to follow up. The requirement of land to raise such seedlings is a small one and does not need government help.

The combined sale of three companies is about 5.75 million seedlings per annum helping the companies to raise revenue of Rs. 72 million per annum. The figure is likely to increase in the future. The success has led to mushrooming of smaller nurseries selling seeds and seedlings, and they may be covering more than 26,000 ha per year with improved clones, mostly of eucalyptus and poplar. Industry is thus making substantial contribution to the production of wood on non-forest lands.

Leasing of Degraded Forest Lands

Degraded forestlands may have a low tree density, but satisfy the fuelwood, fodder and livelihood needs of many poor people, thus they are a vital source of living for the poor. Alienation of forests will, therefore, result in hardships and oppression for the local communities who have historically depended on such lands for meeting their basic needs. The claim of the industry over forestlands is, therefore, against the interest of farmers and forest dwellers.

International Agreements

The Kyoto Protocol was adopted at the Conference of Parties (CoP) of the United Nations Framework Convention on Climate Change (UNFCCC) at Kyoto in 1997, wherein, the developed countries along with some countries falling under category of Economies In Transition (EITs) categorized countries, were required to reduce their emission to less than 5% of 1990 level by the year 2012.

Carbon sequestration is considered as one of the important strategies under Kyoto protocol, through sink enhancement. The Kyoto Protocol potentially offers an opportunity for capital investment and technology transfer for commercial forestry, under the afforestation/reforestation activities of the Clean Development Mechanism (CDM). Carbon sequestration through sink enhancement by way of growing more trees is one of the cost-effective modes of achieving this objective Noble, I., Scholes, R.J. (2001). Sinks and Kyoto Protocol. Climate Policy 1, 5-25.

Industrialized countries are allowed to buy carbon credits developed by carbon sequestration projects through Land-Use, Land-Use Change and Forestry (LULUCF) options in developing countries. Under the Kyoto Protocol, countries need to invest under LULUCF activities while determining their net GHG emissions under allowed allowances. It gives the industries, targeted for emitting GHGs the flexibility to continue using fossil fuels for some time, which will act as a buffer in the process of switching over to more efficient energy source, if they invest in sequestering carbon. It would help to achieve the goal of sustainable development throughout the world by bringing investment from developed world to developing countries. CDM projects in India could also lead to a large positive impact on programmes aimed at forest conservation and regeneration, reclamation of degraded land and socio-economic developemnt of rural communities in addition to global environmental benefits.

The investment by developed countries would enhance the rate of afforestation/reforestation and other additional activities to meet the biomass needs and reclaim degraded and wastelands in the country. Now this is being looked upon as a major benefit with market instruments favouring the active involvement of industry in the mechanism. The potential of India is quite high for CDM related projects. Although it is not mandatory for Indian industries to meet any sort of targets under Kyoto, they still prefer to favour this phenomenon as it strengthens their claim over wastelands of the country.

BARRIERS TO PRODUCTION FORESTRY IN INDIA

Though the expenditure on afforestation by the government in its five-year plans is increasing, still various on-ground problems continue to remain unsolved. There are several reasons, which continue to make adoption of production forestry by Indian farmers and entrepreneurs difficult. Difference in the land ceiling limits in different states needs a mention. The size of land to be held under a single person is different in different states, because land falls under state list. This calls for taking into consideration size of land in each state separately. This makes the case for taking up production forestry on a commercial level a bit difficult.

Financing plays the next important role. Industries are generally interested in financing the leasing of land and other activities for taking up production activities for meeting their raw material requirement by forming direct links with farmers. Schemes such as Investment Promotional Scheme were launched with all good intentions, but failed primarily because of slow progress being made by the financed projects. Forestry, being a long-gestation activity, needs huge investment and failure of such schemes makes the potential financers wary of forestry projects.

The current policies in India favour import of wood for meeting the wood requirements. Exports remain banned. This clearly marks that there is a great need to increase our production forestry if dependence on other countries is to be reduced.

As of now, due to policies favouring the import of wood, our market is open to outsiders. Due to increase in the number of competitors, the profit to small investors is bound to be less. The prime reason is that there are no economies of scale. Instead of importing wood from other countries, the government should facilitate involving local communities and big farmers to take up commercial forestry. This will generate additional employment opportunities besides providing financial incentives for tree growing activities.

The present developments; however, display a ray of hope. India currently is the biggest market for generating Certified Emission Reductions (CERs). The Government of India has set up a National Designated Authority (NDA) to take up CDM projects developed in the country. Technologies for forestry projects are being developed and a few projects have already been approved under CDM. It still remains to be seen how successful these schemes would be. With the demand for raw material expected to increase as a result of growing consumerism and globalization, efforts need to be taken to ensure the success of these schemes. If the response and concern of government is something to go by, then the success of these schemes is very much possible, provided various stakeholders are involved.

3

Changes in Forest Cover and Condition

CHANGES IN FOREST COVER OVER TIME

The periods we studied had quite different effects on the forest coverin El Salvador. We used satellite imagery from before the time of the peace accords in the early 1990s to the early 2000s to quantify changes in forest area. For this period, we used two independent but complementary data sets, derived from (1) high-resolution Landsat Thematic Mapper (TM) imagery, and (2) AVHRR (Advanced Very High Resolution Radiometer) and MODIS (Moderate Resolution Imaging Spectroradiometer) imagery. The data sets were used to map areas of tree cover for each period and changes in those areas as a result of shifts in land use. Before we discuss the results of the analysis, it is important to clarify the definition of woodland in this study. By "woodland," we mean a land area of more than 0.5 hectares (ha), dominated by woody vegetation and with a canopy cover of more than 10%. This definition has been used by the Food and Agriculture Organization of the United Nations, the World Bank, and other institutions in the region, and it provides a measure for quantitative estimates of such forests from remote-sensing data. This also helps avoid the semantically fraught complexity of the unqualified use of the word "forest." Chronosequences of Landsat imagery over many tropical regions are not widely available because of continuous cloud cover and limited data acquisition. In this study, we obtained orthorectified Landsat data for two periods of the early 1990s and 2000s from the Global Land Cover Facility.

CHANGES IN FOREST COVER

A review of the individual requirements of global change, forest inventory, and policy communities reveals a striking convergence, or similarity in scope and definition. The most complex and stringent requirements are set by the forest inventory users since they have needs for detailed forest cover characteristics linked to specific in-situ data with high spatial accuracy and precision. The carbon cycle community and policy users (particularly the emissions inventory requirements) have of subset of these needs. Thus, it is possible to define a single common data bundle which satisfies a very broad range of users and at the same time focuses on the requirements for the global carbon community, forest inventory community, and the policy community.

Between 1980 and 1995, the extent of the world's forests decreased by some 180 million ha, an area about the size of Indonesia or Mexico. This represents a global annual loss of 12 million ha, an area equivalent to the size of Greece or Bangladesh. During this 15-year period, developing countries lost nearly 200 million ha of natural forests, mostly through clearing for agriculture (shifting cultivation, other forms of subsistence agriculture, the establishment of cash crop plantations such as oil palm, and ranching). This was only very partially compensated for by the establishment of new forest plantations. Over the same period, forests in the developed world expanded slowly (by some 20 million ha) through afforestation and reforestation, including natural regrowth on land abandoned by agriculture.

Forest plantation and natural regeneration on abandoned agricultural land more than compensated for clearing of forests due to urbanization and infrastructure development in most industrialized countries, outside the former USSR. While the gain of forest cover in developed countries was 20 million ha in the period 1980-95, 9 million ha of this increase occurred in the 5 years from 1990 to 1995.

The situation is quite different in the developing world, where deforestation exceeded net afforestation/reforestation, particularly in the tropical zone. As a whole, the annual rate of deforestation in the developing world between 1990 and 1995 was 0.7%, equivalent to 12.6 million ha, with the highest rate in tropical Asia-Oceania, closely correlated with population and income growth.

The lowland forest formations were the most affected by deforestation, although the proportional loss during this period was greater in upland formations (1.1% compared with 0.8% in the lowland formations). There is some evidence that the rate of loss of forest cover in developing countries was slowing towards the end of the 15-year period 1980-95. The annual rate of forest loss in the period 1980-90 was 15.5 million ha compared with 13.7 million ha between 1990 and 1995.

IMPACTS ON FOREST GROWTH AND PRODUCTIVITY

Many aspects of projected climate change will likely affect forest growth and productivity. Three examples are described below: increases in carbon dioxide (CO_2), increases in temperature, and changes in precipitation.

- Carbon dioxide is required for photosynthesis, the process by which green plants use sunlight to grow. Given sufficient water and nutrients, increases in atmospheric CO_2 may enable trees to be more productive. Higher future CO_2 levels could benefit forests with fertile soils in the Northeast. However, increased CO_2 may not be as effective in promoting growth in the West and Southeast, where water is limited.
- Warming temperatures could increase the length of the growing season. However, warming could also shift the geographic ranges of some tree species. Habitats of some types of trees are likely to move northward or to higher altitudes. Other species may be at risk locally or regionally if conditions in their current geographic range are no longer suitable. For example, species that currently exist only on mountaintops in some regions may die out as the climate warms since they cannot shift to a higher altitude.
- Climate change will likely increase the risk of drought in some areas and the risk of extreme precipitation and flooding in others. Increased temperatures would alter the timing of snowmelt, affecting the seasonal availability of water. Although many trees are resilient to some degree of drought, increases in temperature could make future droughts more damaging than those experienced in the past. In addition, drought increases wildfire risk, since dry trees and shrubs provide fuel to fires. Drought also reduces trees' ability to produce sap, which protects them from destructive insects such as pine beetles.

CONVERSION OF FORESTS TO OTHER LAND COVER

The Forest Resource Assessment 1990 carried out an assessment of the relative importance of the various factors involved in deforestation at regional and global levels between 1980 and 1990 in the whole tropical belt. Among the most significant outputs of the study were the 'area transition matrices' of the type, which indicate transfers from one land cover class to another over the 3068 million ha of the tropical zone covered by the sample.

For instance, the first row shows that 1275.9 million ha of the total area of 1368 million ha of closed forest in 1980 remained as closed forest up to 1990 and also shows the fate of the 92.1 million ha converted to other land cover classes (open forest; long fallow; fragmented forest; shrubs and short fallow; other, *i.e.* non-wooded, land cover; and forestry or woody forest and agricultural tree plantations). All these transfers involved changes of woody biomass attempts to capture by replacing each class along the *y*-axis (average biomass per hectare) and showing the importance of transfers along the *x*-axis.

RECENT ESTIMATES OF GLOBAL FOREST AREA

A globally consistent methodology using satellite imagery was implemented to quantify gross forest cover loss (GFCL) from 2000 to 2005 and to compare GFCL among biomes, continents, and countries. The following section is based on 1990 baseline figures, updated to 1995, prepared by the Food and Agriculture Organization (FAO) Forest Resource Assessment Programme, reported in and largely drawn from *State of the World's Forests 1997*.

In 1995 forests were estimated to cover 3454 million ha, or 26.6% of the total land area of the world (Greenland and Antarctic excepted). Almost two-thirds of the world's forests were located in seven countries: Russia, Brazil, Canada, USA, China, Indonesia and Zaire; 29 countries had more than half of their land covered by forest, of which 21 were in the tropical belt. However 49 countries, in addition to the many non-forested small island states and territories, had less than 10% of their land covered by forests. Five entire subregions were in this category: North Africa (1.2% of the land area), Near East (1.9%), Temperate Oceania (6.2%), Non-tropical Southern Africa (6.8%) and West Sahelian Africa (7.5%).

Figure: Reduction in global forest area between 15 and 45 percent.

The latest information on the distribution of forest and of forest cover change by ecological zones was provided in 1990; the results of the Forest Resources Assessment 2000 were released by FAO in 2000. In 1990, temperate and boreal forests occupied 1.64 billion ha and tropical forests 1.76 billion ha. A breakdown by ecological zone was available only for tropical forests.The figures on tropical forests show that most (88%) are in lowlands; of these, tropical rain forests accounted for 47% of all lowland tropical forests, followed by moist deciduous forest (38%) and dry and very dry formations (15%).

Estimates of Natural Forests Area

The forest area estimates described above include undisturbed forests, forests modified by humans through use and management (or 'seminatural' forests) and forests created artificially by humankind (*i.e.* forest plantations) by afforestation or reforestation. (Afforestation is defined as the establishment of a tree crop on an area from which it has always, or for a very long time, been absent. Reforestation is defined as the establishment of a tree crop on forest land.)

In most industrialized countries, particularly in continental Europe, forests are being managed in such a way that at management-unit level a continuum exists from low-intensity management, involving natural regeneration, through more intensive methods involving some artificial planting to highly intensive methods with complete planting and cultivation; this makes it difficult to isolate figures for natural forest and plantations. The distinction between natural or seminatural forests and forest plantations can more easily be made for developing countries and some industrialized countries such as New Zealand in which forest plantations have been established using introduced species.

Interest in natural forests, particularly their role in the conservation of biological diversity, has led to efforts to compare forests today with what is thought to be their original character and to give complete protection to areas of forests which have had no, or minimal, human interference. Although there are difficulties in identifying the extent of natural forest, compounded by problems of definition, some information exists that can be used as an indication of broad patterns of natural forests in various regions.

An attempt has been made by the World Wide Fund for Nature (WWF) to quantify the area of forests in western Europe that has been relatively undisturbed by humans or which has retained much of its natural character. Their report distinguishes between 'virgin forest', defined as 'forest ecosystems whose characteristics are determined exclusively by natural location and environmental factors... without human influences present or visible any more', and 'natural and ancient seminatural forests', which 'have not been planted or sown by man for the past two centuries' and 'which continue to have a large number of the natural elements'.

The study found that only a small proportion (probably < 1%) of the total forest land in northern and western Europe could be considered as virgin forest, which has arisen since the last glaciation. Almost all was located in Sweden, Finland and Norway, with small areas in Greece, Austria and Switzerland and (according to another author) in France. In eastern Europe, Slovakia and Belarus have considerable areas of virgin forest while Poland

and Croatia have small areas. In addition, in northern and western Europe, the WWF report identifies natural and ancient seminatural forests representing 2.1% of the total forest cover (1990) of the 16 countries concerned. There are a further 3 million ha of land in the region in national parks and other protected areas and another 50 000 ha in small forest reserves (mostly for nature conservation and scientific research), whose use is tightly restricted.

The situation in temperate and boreal North America is quite different from that of densely populated Europe and Japan, where use and management of forests for many centuries have left very little of the original forest area untouched. 'Old growth forests', as they are called in North America, still cover extensive areas. On the lands managed by the US National Park Service alone, old-growth forests covered 1.97 million ha in 1988. Although not a direct measure of the extent of old growth forest, it may be noted that the area of forest included in national parks and other protected areas in North America (USA and Canada) was reported to be nearly 49 million ha in 1990.

The FAO Forest Resource Assessment 1990 did not distinguish between undisturbed and disturbed 'natural' forests in developing countries. However, the Forest Resource Assessment 1980 made estimates of the areas of undisturbed closed forests (primary forests and old secondary forests where there had been no logging for the last 60-80 years) and of closed forests that were included in national parks and other protected areas (thus relatively undisturbed, at least in theory).

At that time, these two categories together represented 60% of the total closed forest area in the tropics, a proportion varying from 39% in tropical Asia to 59% in tropical Africa and 69% in tropical America. These different proportions by region reflected a slower development of large-scale harvesting in tropical America compared with tropical Africa and Asia, and also the fact that, in tropical America, spontaneous colonization did not follow in the wake of logging as systematically as in the two other regions due to lower population pressure. Although the two categories do not match the concept of 'virgin forests' of Europe and of 'old growth forests' of North America discussed above, the sum of the two categories nevertheless gives an indication of the amount of forest disturbance (or management, depending on the point of view of the observer) that existed around 1980 in the humid tropics. Although corresponding estimates for 1990 and 1995 are not available, it is likely that the share of undisturbed forests remains higher in the three tropical regions than in Europe and probably in North America too.

HISTORY OF FOREST PLANTATIONS

By the seventeenth century the decline in the area of native forest in European countries led to the planting of trees, largely to provide alternative sources of timber supply. Sometimes plantations were established for the provision of services or other products than timber. They might be planted as shelterbelts, for dune stabilization, for amenity or for the supply of firewood.

Plantations of native species were established at first in areas where forests occurred naturally, for example France, Germany, England and Scotland, but later also on previously unforested land. In France, for example, planting of *Pinus pinaster* was started on the sand dunes of the Landes and with *Pinus sylvestris* and *Picea abies* on former agricultural land in the Vosges; in Germany likewise agricultural land was planted with Norway spruce *(Picea abies)* in Saxony. In tropical countries such as Myanmar (then Burma), teak *(Tectona grandis)* was planted as a native species in the *taungya* system.

Teak was one of the first exotic forest plantation species to be used, being planted in Sri Lanka and the island of Java (Indonesia) from early in the nineteenth century. Exotic species, such as Douglas fir *(Pseudotsuga menziesii),* Sitka spruce *(Picea sitchensis),* Japanese larch *(Larix kaempferi),* lodgepole pine *(Pinus contorta)* and poplar *(Populus deltoides),* were introduced to Europe during the nineteenth century, when they began to play an increasingly important role in forest plantation programmes. Other exotic species, particularly the eucalypts *(Eucalyptus* spp.) and wattle *(Acacia* spp.) from Australia, were introduced as exotics in tropical and subtropical countries from the middle of the nineteenth century, while *Pinus radiata* was introduced from California to New Zealand and other countries such as Chile from the early years of the twentieth century. Box 1.3 describes in detail the experience with forest plantations in Denmark.

The main reason for the establishment of plantations remained the decline in natural forest area and a scarcity of wood and, the conversion of woods and forests to agricultural and grazing use. In recent years increased areas of natural forest have been managed for nature conservation, recreation, wildlife parks, etc., and commercial wood production has been either reduced or eliminated. For example, over 25 000 ha of high-yielding hybrid poplar plantations have been established in the north-western USA between 1992 and 1997 in response to both increased demand for poplar wood for orientated-strand board and decreased supply from public forests. On the other hand, some new areas have become available in European countries as land is taken out of agriculture due to trade and market considerations.

Tree Crops in Forest Plantations

Forest plantations are tree crops that are in some, but not all, ways analogous to agricultural crops. They often have a simple structure, at least in youth, and are usually composed of one or a few species (but not varieties as in agriculture, except in a few cases such as the intensively bred poplars) chosen for their fast growth, yield of specified products and ease of management. When established for wood production they have a higher productivity of usable wood than natural forests, but due to the way they are managed they do not, indeed cannot, provide the full range of goods and services that natural, seminatural or even secondary forest can provide.

Although the objective of many plantations is the production of industrial roundwood and/or fuelwood, many are established for environmental protection or other services (*e.g.* soil and water conservation, enhancement of agricultural production in agroforestry systems, carbon sequestration) and some also provide non-wood forest products, such as fodder, various foodstuffs, medicines, etc.

Area of Forest Plantations

The area of forest plantations throughout the world started to increase in the 1970s as many governments became concerned about wood supplies for industry, and in developing countries fuelwood supplies, and has continued to increase since. However, there are no reliable global figures for plantation areas because forests of native species in several developed countries in the temperate and boreal regions, especially in continental Europe, are frequently regenerated naturally and it is not possible to distinguish those areas where supplementary artificial planting has been done. Furthermore many countries consider their plantations as 'seminatural' forests over a certain age, because as stands mature the clear initial row layout of the trees is lost and other species naturally regenerate under the canopy. For example, Austria, Czech Republic and Finland, in responding to a questionnaire to collect data for the ECE/FAO Temperate and Boreal Forest Resource Assessment (TBFRA), stated that they had no plantations in their countries as defined by the TBFRA process.

In 1995, an approximate estimate of the area of plantations in developed countries was 60 million ha, comprising 13.7 million ha in North America, 22.2 million ha in the Commonwealth of Independent States, 12.1 million ha in Europe and 13.2 million ha in Oceania (Australia, New Zealand, Japan). The most significant areas of plantations were in the Russian Federation (17.3 million ha, 2.1% of the country's total forest area), the USA (13.7 million ha, 6.3%), Japan (10.7 million ha, 44.4%), Ukraine (4.4 million ha, 46.8%) (FAO/ECE 1996), Spain (1.9 million ha, 14.9%), New Zealand (1.54 million ha, 19% of the total forest area in 1996, 91% of which was *Pinus radiata)* and Australia (1.04 million ha in 1994, 85% of which was softwood species, mainly *P. radiata).*

The estimated 'net' plantation area of 55 million ha in developing countries in 1995 was about 2.8% of the total area of forests in developing countries. In 1980 the net plantation area was assessed at about 40 million ha. It has thus increased by about 15 million ha in 15 years, but even this figure may be liable to error. Many developing countries, especially those with large forest plantation programmes, provided updated information on their present plantation plans to FAO in 1996 and 1997 from which the reported annual rate of new plantations of 3 million ha was derived. Note, however, that this is the reported or planned rate, which may not necessarily have been achieved.

Most of the countries with large plantation estates indicated that they intended to double their plantation areas between 1995 and 2010.

It was estimated from reported figures that 57% of the forest plantation area consisted of hardwood species and 63% was established for industrial purposes. Nearly three-quarters of these plantations were in the Asia-Pacific region, where China (21 million ha) and India (20 million ha) dominate, while about 15% were in Latin America and 10% in Africa.

One of the trends in the tropics has been that the proportion of industrial plantations established in large blocks fell from 40% of the total plantation area in 1980 to 35% in 1990. The proportion of smaller plantations established through farm forestry or agroforestry programmes grew in importance during the period 1980-95, particularly in the Asia-Oceania region. Unfortunately, the figures on small-scale private or community-owned forest plantations are even less reliable than for large-scale plantations. Some of these farm forestry or agroforestry plantations supply industrial wood markets for pulpwood. In many countries, particularly those with limited forest area, planted trees grown outside the formal forest area often provide the bulk of fuelwood, poles, construction wood, utility wood, as well as fodder and other non-wood forest products for household use.

Area of Hardwood Plantations

Of the area of hardwood plantations planted for industrial use, 30% or nearly 10 million ha consists of eucalyptus, followed by acacias (3.9 million ha or about 12% of the hardwood area) and teak (about 7%). Short rotation plantations of hardwood species such as the eucalypts, acacias and *Gmelina arborea* have been grown for many years by the private sector, but the establishment of plantations of teak or other valuable hardwood species has been carried out only by government forest services because of their slow growth and hence delayed returns. However, the likelihood of reduced supplies of high-quality hardwood logs derived from natural forests, combined with increasing purchasing power and expected higher prices for logs, is leading to increasing interest in investment by the private sector in valuable hardwood species, especially teak, in a number of countries, for example India, Malaysia, Costa Rica and Ghana.

Of the softwoods grown for industrial purposes, fast-growing pines such as *Pinus radiata, P. patala and P. caribaea* constitute about 25% of the area while other, often slow-growing pines make up about 36%.

Responsibility for the monitoring and regulation of plantation crops grown for food and certain other purposes has long been the job of the agricultural sector. In recent years, however, several of these crops (the main ones being rubber, coconut and oil palm) have been providing 'forest' products that have been used for wood and fibre. For example, rubber wood is now used for the manufacture of about 80% of the furniture made in Malaysia, while coconut and oil palm trunks and the branches of rubber wood are used for various forms of reconstituted 'wood'. Rubber wood and coconut stems are derived from the conversion of old plantations formerly disposed of by burning, while oil palm fruit residues are used for medium-density fibre board. The development of these new markets has thus not only improved financial returns but also used the resources in an environmentally friendly manner.

The area of these species appears to have increased from the 14 million ha reported in 1990. The increase may be due to better coverage of the data, although it is known that oil palm areas are increasing rapidly, rubber tree areas are also increasing and coconut plantations are decreasing. Coconut plantations comprise the largest area (about 42% of the total), rubber 36% and oil palm 22%. Most of the coconut plantations are in Indonesia (33% of the area) and the Philippines (28%); most of the rubber plantations are in Indonesia (34%), Thailand (20%) and Malaysia (18%), while most of the oil palm plantations are in Malaysia (44%) and India (29%). Not all of the areas mentioned above are suitable or available for substitute 'timber' production but they illustrate the potential.

CONTRIBUTIONS OF FOREST PLANTATIONS TO WOOD SUPPLY

The continuing increase in the area of forest plantations which have been established for industrial wood supply has been to meet the reduction of outturn foreseen from natural forests arising from deforestation and changes in land use (largely in the tropics and subtropics) or from natural forest being taken out of production and devoted to service functions such as conservation. It has been believed that the outputs from forest plantations can help to reduce the pressure on natural forests as sources of industrial wood supply; while the logging of tropical natural forests is not the prime cause of deforestation, logging roads often provide the means for farmers to gain access to forests. The reduction of logging, combined with effective protection, may thus help to reduce deforestation in certain locations until land use and ownership are clarified. However, none of these palliatives will remove the underlying causes of deforestation: high rates of population growth, poverty, hunger and a shortage of fertile land to cultivate.

The potential of forest plantations to meet demand for industrial roundwood is considerable; it has been estimated that the present global demand for paper pulp could be met from an area equivalent to only 1.5% of the world's closed forest area (IIED 1996). No global estimates of current output of timber from forest plantations are available, although FAO's global fibre supply model (GFSM) estimated that the potential annual growth of industrial wood from forest plantations in developing countries was about 5% of the increment of natural forests in 1995. In some countries, plantation production already makes a highly significant contribution to the industrial wood supply, for example in New Zealand 99% of industrial roundwood in 1997 was grown in plantations, while in Chile the equivalent figure was 95%, in Brazil and Argentina 60%, and in Zambia and Zimbabwe 50%.

Estimating the future contribution of forest plantations to wood supply is at present imprecise and is based on many more or less unreliable assumptions, particularly concerning the rate at which afforestation will continue. By the year 2010 the GFSM (op. cit.) estimated that the potential increment from forest plantations would be about 40% of that from natural forests in Asia, Oceania and Latin America and about 15% in Africa, under rates of deforestation and afforestation largely the same as today.

INDIA'S FOREST COVER DECLINES

India has lost 367 square kilometres of forest cover in the past two years. According to the India State of Forest Report, 2011, released by the Forest Survey of India (FSI) on February 7, the total forest cover in the country is now at 6,92,027 sq km. This accounts for 21.05 per cent of the total geographical area of India.

ANDHRA LOSES, PUNJAB GAINS

- Andhra Pradesh has lost the maximum forest cover—281 sq km—as compared to 2009
- Northeastern states saw an unprecedented loss in forests this year. The region which accounts for nearly one-fourth forest cover of the country has seen a decrease of 549 sq km of forests
- The other states that lost forest cover are Kerala (24 sq km), Chhattisgarh (4 sq km), Maharashtra (4 sq km), Uttar Pradesh (3 sq km), Gujarat (1 sq km) and Chandigarh (0.22 sq km)
- Punjab registered maximum growth of 100 sq km forest cover followed by Jharkhand (83 sq km), Tamil Nadu (74 sq km), Andaman and Nicobar Islands (62 sq km), Rajasthan (51 sq km), Odisha (48 sq km) and Bihar (41 sq km)
- Other states that registered forest growth include Haryana, Himachal Pradesh, Karnataka, Goa, Jammu & Kashmir, Uttarakhand and West Bengal
- A total of 548 sq km forest cover has decreased in the 124 hill districts of the country
- In the 188 districts of the country dominated by tribal population forest cover has decreased by 679 sq km
- The mangrove cover in the country has increased by 23.34 sq km. They are now spread over an area of 4661.6 sq km
- The total growing stock of India's forests and trees is now 6,047.15 million cubic metre which comprises of 4,498.73 million cubic metre

of growing stock inside the forests and 1,548.42 million cubic metre outside the forests

- FSI has for the first time estimated the growing stock of bamboo.
- Total bamboo bearing area in the country is 13.96 million hectare (ha).
- Arunachal Pradesh has maximum bamboo-bearing area of 1.6 m ha followed by Madhya Pradesh (1.3 m ha), Maharashtra (1.1 m ha) and Odisha (1.05 m ha).

MORE DENSE FORESTS

- The dense forest cover (lands with tree canopy density of 70 per cent and above) increased by 43 sq km since 2009 and stand at 83,471 sq km.
- The dense forests now account for 2.54 per cent of the geographical area of the country
- The moderately dense forest (lands with tree canopy density between 40 per cent and 70 per cent) now occupy 320,736 sq km and account for 9.76 per cent geographical area of the country
- The moderately dense forest saw an increase of 498 sq km
- The area covered by open forests (lands with tree canopy density between 1O per cent and 40 per cent) saw a decrease of 908 sq km and stand at 287,820 sq km.

VAGUE REASONS

- In Andhra Pradesh, the survey says, the forest cover has decreased due to harvesting of short rotation crops followed by new plantation and forest clearance in the encroached areas
- In the Northeast, the decrease in forest cover has been attributed to shortening of shifting cultivation cycle and biotic pressure
- The reason for increase in forest cover in Punjab, Odisha, Rajasthan, Jharkhand, Bihar and Tamil Nadu is enhanced plantation in and outside forests and effective protection measures.

FORESTS: THE EARTH'S LUNGS

According to the FAO Global Forest Resources Assesment 2005, the net loss in forest area at the global level during the 1990s was an estimated 94 million ha - an area larger than Venezuela and equivalent to 2.4 per cent of the world's total forests. This was a combination of an annual loss of 12.5 million ha of natural forests and an annual gain of 3.1 million ha in the form of forest plantations.

The 2005 FAO Assessment said that between 2000-2005, the world suffered a net loss of a further 37 million hectares (91 million acres) of forest. Nearly 4 billion hectares of forest cover the earth's surface, roughly 30 per cent of its total land area. Though extensive, the world's forests have shrunk by some 40 per cent since agriculture began 11,000 years ago. Three-quarters of this loss occurred in the last two centuries as land was cleared to make way for farms and to meet demand for wood.

However, global statistics tend to obscure significant differences in forest cover change among regions and countries. Net deforestation rates were highest in West Africa and South America. This was followed by Asia, particularly in South-East Asia, though it was significantly offset by forest plantation establishment in other parts of the continent.

In contrast, the forest cover in the other regions, largely in industrialized temperate countries, increased slightly primarily as a result of natural forest succession on abandoned agricultural land.

The countries with the highest loss of forest area between 1990 and 2000 include Brazil, Indonesia, Sudan, Zambia, Mexico and the Democratic Republic of Congo. Those with the highest net gain of forest area during this period were China, USA, Belarus, Kazakhstan and the Russian Federation.

Along with the physical coverage or quantity of forests, it is equally important to consider the quality of forests.

Quantity of forests (i.e. forest area) alone is an inadequate indicator of the health of a forest ecosystem since much of the world's forests are highly fragmented and face considerable human pressure.

Although deforestation is widely recognised as a major conservation challenge, the related issue of habitat fragmentation receives comparatively less attention. As human pressures increase in both temperate and tropical forests, areas that were once continuously forested have become more fragmented. In the Brazilian Amazon alone, the area of forest that are fragmented or prone to edge effects is 150 per cent greater than the area that has actually been deforested (FAO, 2003).

Small fragments have very different ecosystem characteristics from larger areas of forest, containing more light-loving species, more trees with wind- or water-dispersed seeds or fruits, and relatively few under-storey species. The smaller fragments also have more fallen trees, a more irregular canopy, more weedy species and unusually abundant vines, lianas and bamboos. Thus, they preserve only a highly biased subset of the original flora and fauna, which is adapted to these conditions.

In a 1997 study, the World Resources Institute coined the term "frontier forests" to describe forested areas that are relatively undisturbed by human activity and are large enough to maintain their original biodiversity, including viable populations of wide-ranging species.

According to this study, frontier forests constitute about 40 per cent of total global forest area, but are heavily concentrated in only three large blocks - two areas of northern (boreal) forest (in Canada, Alaska, and Russia), and one relatively contiguous area of tropical forest spanning the northwestern

Amazon Basin and Guyana Shield (in Brazil, Peru, Venezuela, and Colombia). Additional important outliers can still be found in Central Africa (Congo), and Papua New Guinea.

However, nearly 40 per cent of these remaining frontier forests face a moderate to high threat of degradation or clearance.

The demise of frontier forests is a major reason for an alarming loss of biodiversity around the world. The term 'biodiversity,' short for 'biological diversity,' refers to the total variety of all living things on earth, including individual species and whole ecosystems.

4

Forest Climate Surfaces

CLIMATE SURFACES

Climate surfaces that can provide point estimates of climate measures are needed for studying forest plant-climate relationships and for making predictions about how forests, plant communities and species distributions may change on natural landscapes. To meet those needs, Rehfeldt (2006) developed monthly climatic surfaces using Hutchinson's thin-plate splines from which were derived variables of demonstrated importance in biology. Analyses of data have shown that climate estimates produced using these methods are closely related to plant responses and should have a variety of uses in spatial biological research. This web page provides access to these climate data and surfaces, and to some of the predictions we have made using these data. For climate data, we build contemporary climate and future climate surfaces. The future climate surfaces are based on the same data used to build the contemporay surfaces after they are updated using information from General Circulation Models (GCM's). Plant-climate relationships are based on contemporary observations of species presence and absence and contemporary climate. Predictions using these relationships can be mapped for contemporary forests and future forest climates.

The potential migration of treelines in response to changes in global climate may also significantly change the regional exchange of C as CO_2 and CH_4. In the high latitudes, the peatlands have steadily accumulated C since the last ice age but at modest rates relative to net C sequestration by young forests. However, if forests migrate into peatlands, the effective lowering of the water table will suppress CH_4 emission and lead to oxidation of the surface organic layers to CO_2. While supporting forest growth, such changes may represent a large net source of CO_2 in the atmosphere over substantial areas, as currently observed when peat wetlands are drained and planted with trees. The extent to which these changes may occur at the northern boundary of the boreal forest during the coming decades will be controlled by the changes in climate at these high latitudes, and on the balance between precipitation, and evapotranspiration and the length of growing season. At present the possible role of changes in nutrient inputs from the atmosphere remain very uncertain at these remote locations and the major controls are likely to be driven by climate.

It came as a considerable surprise when Bowen ratios with values of 2 or more for coniferous forest, measured by flux-gradient methods, were first reported in the 1960s because virtually all previous measurements over agricultural crops had resulted in values between about -0.5 and +1.0, except in a few cases of extreme water stress. Suspicion was placed on the methods being used, particularly on the problems of measuring with sufficient accuracy over sustained periods the very small gradients of temperature and humidity that occur in the surface layer over forests and which are required for the so-called Bowen ratio, flux-gradient method.

For those who had worked previously only with prairie grasslands and crops, it still came as a surprise 25 years later to discover that values of p measured by eddy covariance for stands of black spruce and jack pine, which comprise the most extensive areas of the Canadian boreal forest, were frequently larger than 2, and this has revolutionized thinking with respect to climatic process at larger scales in the boreal region. The significance of a large Bowen ratio at regional scale lies not so much in the transpiration rate but in the large sensible heat flux which, in extensive continental areas drives the CBL up to heights of 3000 m during just an hour or two in the morning, with consequent entrainment of cold dry air rich in CO_2 from above the capping inversion which, as it warms up, strongly forces evaporation and transpiration and replenishes CO_2 over cropping areas as well as the forest within the region.

CLIMATE

Climate is the dominant control factor in the distribution of the world's main vegetation types or biomes. Biomes are defined as a distinctive ecological system, characterized primarily by the nature of their vegetation. On a global scale and in a historical and geological time frame, response to climatic stress is of overriding importance. The relevance of understanding tree and forest ecosystem response to climatic stress has increased because we are currently experiencing an unprecedented change of global climate which is, at least in part, the result of anthropogenic factors.

Most contemporary texts on the influence of climate and vegetation date our understanding of the relationship between climate and the distribution of vegetation from the work of Von Humboldt at the beginning of the nineteenth century. Holdridge (1947 and subsequent work) demonstrated a strong correlation between the main groups of life form or biomes (*e.g.* deserts, grasslands, deciduous or coniferous forests) and two major climatic characteristics, temperature and water availability. These climatic factors are the main influence at the biome scale, while soil type (nutrient availability) topography and anthropogenic factors are major influences on vegetation patterns within biomes on a smaller local scale.

It is of considerable interest and relevance that where temperature and moisture availability or moisture budget (defined as precipitation-evapotranspiration) are similar, the same tree species or similar species of one tree genera are present or prove to be satisfactory forestry species if introduced. This reflects the relationship between climate and suitable physiology and anatomy, and also the pattern of species' evolution and migration. The timing of the separation of the earth's present-day continents relative to the evolutionary history of tree species and taxa determines presence or absence of particular species within suitable climatic zones. Eventually the land bridges between the continents became impassable to tree species during the period of their rapid evolution in the Cretaceous Era. In the absence of human influence, the geographical extent (natural distribution) of the various forest biomes is thus determined mainly by climatic conditions and the natural species composition will be further influenced by evolutionary and geological factors.

There are several critically important temperature thresholds that affect physiological processes and thus define the environmental limits of the major forest biomes. The first threshold is a critical minimum temperature in the range +12 to 1°C, which results in chilling injury to plant membranes. The next critical threshold is at about -15°C and roughly corresponds to the temperature at which frost damage occurs to leaves and buds (depending on the degree of frost acclimation).

The leaves of broadleaved species with a damage threshold at about -15°C are less resistant to frost than coniferous needles. The buds of broadleaved and coniferous species have similar minimum temperature thresholds but the xylem of broadleaved trees (angiosperms) is more susceptible to ice formation than that of conifers. This is because coniferous xylem is made up of tracheids, which have a smaller lumen diameter than the vessels that make up the xylem in broadleaved trees. Similarly broadleaved species with diffuse-porous xylem, which is of smaller diameter than ring-porous xylem, are more resistant to ice-induced xylem damage. Broadleaved trees with ring-porous xylem are notably absent from boreal forests.

The next critical temperature threshold is at -39 to -40°C, below which only a few boreal tree species can survive (*e.g. Picea glauca, Larix laricina* and *Larix sibirica)* because this temperature will freeze the xylem and buds. At their extreme northern limits, these cold-tolerant species often have a parenchyma pith cavity beneath the crown of the primordial bud that prevents ice formation from spreading from the xylem to the bud.

It has been argued that in harsh climates, particularly low temperatures, the evergreen habit gives a more favourable carbon balance and thus we see the dominance of coniferous evergreens in cold montane and boreal forests. Climates with ample rainfall, well distributed throughout the year and with rare frosts, as experienced in temperate regions (30-55° of latitude), also favour evergreen rather than deciduous forests. This is considered to have led to the dominance of broadleaved evergreen forests in the temperate regions of the Southern Hemisphere, with the notable exception of southern Chile where deciduous broadleaved forests occur. Broadleaved deciduous tree species predominate in the northern temperate regions. The distribution of the various forest biomes is thus determined by climatic conditions. However within the major biomes individual species gradually decrease in numbers near their altitudinal or latitudinal limits. Woodward attributed this more to increasingly non-competitive carbon balance rather than to direct damage from low temperatures. Although the tree species present within the various forest biomes should not normally be experiencing climatic stress, climate remains of major importance to forestry and to the conservation of forest ecosystems for three main reasons. First, within a climatic region foresters are able to select species that grow better (faster) or provide other desirable benefits for particular sites, as influenced by local conditions of topography and soils. Second, the effect of climate on tree species limits the use of introduced or exotic species, which otherwise has proved a very successful strategy for improving yields in commercial forestry worldwide. Third, and perhaps of greatest interest, is that climate changes, not only naturally on a geological time-scale but also rapidly as a result of human activity, *i.e.* global climate change.

Landsberg and Gower (1997) followed Melillo *et al.* (1993) in adopting a classification of forest biomes based on major climatic region (tropical, temperate and boreal) and tree physiognomy (broadleaved, evergreen, broadleaved deciduous and needle-leaved evergreen conifer). Seven forest biomes were identified as follows: boreal, temperate deciduous, temperate coniferous, temperate mixed, temperate broadleaved evergreen forest, tropical evergreen and tropical deciduous. Of these, the boreal and tropical evergreens cover the greatest area.

The increase of global surface land and ocean temperatures that has been observed during the last 100 years is faster than that observed during the geological past. Predictions of future climate indicate that some tree species, and some parts of the forest biomes, will experience climates outside those of their natural range. Over significant areas climatic stress will occur and dieback is likely. A number of authors have used existing relationships (correlations) between climatic factors and biome distribution to predict the future distribution of terrestrial biomes from the climate predictions provided by general circulation models. Changes in biome distribution might be expected to occur by migration, although migration at rates of 5 km annually will be required and palaeoecological records indicate postglacial migration rates of tree species of only 40-150 m annually. Thus migration rates are likely to be insufficient, resulting in significant areas of stress and dieback, or rather, that stress and dieback will be part of the process of migration. A recent example of such predictions is that of Friend *et al.* (1997) who mapped the global distribution of eight biomes and of desert for existing (today's) climate and for climate predicted for the year 2080.

This model, called 'HYBRID', predicts vegetation type at any particular location on the basis of the outcome of competition for light, nitrogen and water and of the plants' ability to survive extreme conditions. Unlike earlier, less sophisticated work of this type, HYBRID predicts that tropical forests will decline in area, with dieback beginning in the 2050s as a result of decreased rainfall and increased temperatures. Temperate and mixed forests are likely to move northwards and increase their total area. Boreal forest area is predicted to change very little provided that these forests can establish in more northern areas with shallow poorly developed soils. The palaeoecological records show that the species composition of forest ecosystems varied during their postglacial advance so we could also predict some changes of species composition as a result of the stress caused by climate change.

These predictions suggest that forest managers will need to plan for, and respond to, climate change if extensive areas of forest dieback are to be avoided; some such problems are likely to occur in any event. Indeed, there is firm evidence that climatic conditions have been a major, or perhaps *the* major, factor responsible for forest dieback in Europe and the north-eastern USA. Changes of species choice will be an obvious course of action in plantation forestry. The successful use of 'exotics' in commercial forestry, for example Sitka spruce *(Picea sitchensis)* in western Europe, *Pinus radiata* in New Zealand and Chile, and *Eucalyptus* in Portugal and South America, indicates that forest managers can, in any event, achieve increased yields by planting species outside their current natural distribution.

Species choice and the use of the 'right tree for the site' has long been recognized as a very effective mechanism of avoiding stress to plantations and of optimizing yield. Various systems of biogeoclimatic ecosystem classification (BEC) have recently been adopted in different parts of the world. Essentially such systems provide site classification based on our understanding of the influence of environmental factors on forest ecosystems and thus allow sound management of plantation and seminatural stands for timber production, conservation or other benefits.

The basis of BEC, and the UK equivalent, ecological site classification (ESC), is consistent with our understanding of the climatic control of biome distribution and the influence of other factors, particularly nutrient availability (but also topography and windiness), on species composition within the major biomes. Sites are classed on the basis of three principal factors considered to control site quality: climate, soil moisture regime (on which climate exerts a significant influence) and soil nutrient regime. Choice of species is arguably the most important management and silvicultural decision. ESC provides a list of species suited to the climatic and soil qualities and capable of producing good-quality timber. Tree species are listed as *optimal* if they will grow at the upper rate given by models of general yield class (GYC) without undue risk of disease and pest attack. Species are rated as *suitable* if they will grow on the lower part of GYC curves. Where ESC predicts increased risk of injury from pests, diseases or drought species are considered *unsuitable*. Predictions of GYC can be made for a particular species.

Growth rates and productivity are influenced by climatic stress, although current indications are of increased growth prior to dieback, perhaps even prior to catastrophic decline. The important study by Spiecker *et al* (1996) indicated that the growth rates and timber yields of European forests are currently increasing relative to those achieved in earlier rotations. This is probably the result of better silviculture and planting stock, the use of fertilizers and suitable thinning regimes. However, the effects of warmer weather, nitrogen depositions and elevated CO_2 have almost certainly also contributed to this effect. Perhaps hard to believe is the prediction of the HYBRID model that the potential biomass between latitudes 30° and 60° N will increase by 70% by the year 2080 *cet. par.*

Biogeoclimate classifications can be developed further to provide advice on other aspects of management, in particular risk from windthrow, ground preparation, regeneration techniques and management of open space. Currently, historical climatic data (weather records) are used to 'par-amaterize' these systems of site classification; however, consideration of rotation length relative to the predicted rates of climate change indicates that the use of predicted climate of the site might provide more accurate advice on species choice and future management. Over a number of years, the evidence from average global surface temperatures has become increasingly clear and 1997 was the hottest year on record to date.

Average global temperatures have risen by 0.6°C in the last 130 years and general circulation models are now able to predict the impact of increasing greenhouse gas concentrations on global temperatures. It is now accepted that anthropogenic climate change is occurring, with the UN Intergovernmental Panel on Climate Change (IPCC) concluding in its 1995 report that 'the balance of evidence suggests a discernible human influence on global climate'. To understand the potential impacts of climate change on forests it is essential to have accurate predictions of future climate at the regional scale. Internationally, predictions or scenarios have been provided by the IPCC; within the UK, they have been provided by the Meteorological Office based at the Hadley Centre, Reading, at a spatial resolution of 10min (*i.e.* 10 km grid squares). Mean winter temperatures and average precipitation (mm daily) are available for the 1961-90 baseline period and for the years 2025, 2050 and 2100. Relatively small changes in temperature (+1.0-1.8°C) and rainfall (-9% in summer to +10% in winter) will lead to much larger changes in potential evapotranspira-tion (40% increase predicted for southern England by 2050), while the annual distribution of rainfall and the occurrence of storm events may also change. Projected temperature increases should therefore not be viewed in isolation. The changes predicted for the UK would have very important implications for forestry.

The occurrence of late spring frost is already a problem in forestry and restricts the use of some tree species and provenances. Similarly, increased windiness and particularly storms could have devastating effects. Two important features in forestry are that a number of introduced species are used because they grow rapidly over short rotation lengths and, secondly, that native species have ranges which span considerable climate differences. Seed origins (provenance) are used to select the most appropriate genotype to maximize growth potential. Therefore, sourcing seeds from different origins is a possible proactive response to climate change: for example for Sitka spruce it is suggested that more Oregon or Washington provenances should be planted in place of those from the more northerly Queen Charlotte Islands. Similarly, species choice could be adjusted, with species such as *Eucalyptus* and *Nothofagus,* natives of regions in warmer climates, becoming suitable for Europe. In addition to temperature and extremes of climate (unusually late spring frosts, storms and drought), there are a number of other environmental factors that are changing as a result of human activities. These include UV-B radiation and air pollutant concentrations and depositions. Such factors and their interacting effects on trees and forest insects and pathogens are likely to have significant effects on tree growth and forest productivity.

DYNAMICS OF CARBON DIOXIDE EXCHANGE

DAILY DYNAMICS

Over the course of a day, C sequestration essentially follows solar radiation (provided there is no major constraint such as frozen or dry soil), with C being accumulated during the daylight hours and lost at night. For example, large daytime gains by tropical forest stands are offset by large night-time losses. Depending on the balance between these hourly gains and losses, for any stand there may be a net gain or loss of C over any 24 h in the year. Typically, through a growing season, there may be a net gain of C on about three-quarters of the days but a loss on days of particularly low solar radiation input (*i.e.* with cloud and rain, or smoke in the atmosphere) and high temperatures.

Seasonal Dynamics

Over a year, the length of the photosynthetic season has a major influence on C transfer. In boreal coniferous forest, for example, one-third to a half of the C gained in the five summer months is lost by respiration over the seven winter months, during which a small efflux of C continues although the ground is frozen. In broadleaved temperate forest the length of the photosynthetic season is determined by budburst in the spring and leaf senescence in the autumn, whereas in temperate coniferous forest the season of net photosynthetic gain is determined by daylength and daily total radiation input in the winter months and may be 10 months long for coniferous forest in maritime climates. In boreal, temperate and Mediterranean climates the vegetation is strongly seasonal and NEE may be strongly constrained by water availability for a large part of the time; only in tropical moist forests is net C gain near continuous throughout the year, being reduced by occasional short periods of low temperature and moderate seasonal water shortage.

CARBON DIOXIDE FROM THE ATMOSPHERE TO FORESTS

TROPICAL FORESTS

So far there are few C flux measurements over periods long enough to give annual estimates of NEE in tropical forests, although they are likely to increase appreciably over the next 5 years with the onset of the LBA experiment. Eddy covariance measurements indicate an annual net C sink of about 1.0 Mg ha^{-1} in pristine, sparse, dry, rain forests in Amazonia and about 5.5 Mg ha^{-1} for dense, moist, rain forest.

Temperate Forests

Forests in the temperate region are virtually all managed to a greater or lesser extent and there are but few patches that might be regarded as pristine. Recent measurements of NEE by eddy covariance over a year or more in Europe in the EUROFLUX experiment and in North America indicate rates of C sequestration in the range 1.5 to 7 Mg ha^{-1} over about 20 sites differing in latitude, altitude, climate (continental vs. maritime), species, sparseness of cover and degree of management. The values at the lower end of the range derive from sparse sites where lack of water is a major factor or from northern sites where length of growing season, lack of nutrients and low temperatures are major factors. Estimates of NPP of sample plots indicate annual carbon sequestration of approximately 3 Mg ha^{-1} (range 1 to 8 Mg ha^{-1}) based on numerous yield tables. A preliminary scaling-up exercise of the EUROFLUX measurements suggests that the largely temperate forests of the European Union are a current annual C sink of 0.17 to 0.35 Gt.

Boreal Forests

The extensive boreal forests across Siberia and Canada are fire-driven climax forests, the near end-point of succession following retreat of the ice after the last glaciation about 10 000 years ago. These forests have been subject to increasing exploitation over the past 150 years but much of the area standing can be regarded as largely pristine. Measurements of C sequestration by eddy covariance on a campaign basis in Siberia, over periods of several months up to 5 years in northern Canada, in the BOREAS experiment and in northern Europe have demonstrated old-growth spruce stands that are C neutral in Canada and a C source in Sweden, losing C at an annual rate of up to 1 Mg ha^{-1}. However, the majority of boreal forest stands investigated, some 12 or more, are C sinks sequestering C at annual rates of up to 2.5 Mg ha^{-1}. Estimates of NPP from sample plots also indicate that boreal forests are weak C sinks of about 2.5 Mg ha^{-1}.

Carbon Balance Components

Carbon Stocks and Flows in Forest Biomes

To understand the magnitude of the net transfer of CO_2 between stands and the atmosphere at a particular stage in a succession and to appreciate its sensitivity to current conditions, disturbance and climate change, it is helpful to consider the principal component processes within the stand that add up to give the overall net exchange. The feedbacks among these processes, their speed of response and their sensitivity to environmental change determine the present and future flows of CO_2 and the C sequestration capacity of the stand.

It is evident that the annual NEE is the difference between a very large gain term (*i.e.* photosynthesis of the foliage) and a very large loss term dominated by respiration of heterotrophs in the soil. Thus the net exchange is particularly sensitive to relatively small proportional changes in the opposed gain and loss terms, particularly if they are out of phase.

A mass balance of the component fluxes above ground (photosynthesis; foliage, branch and stem respiration; leaf, branch and stem litter of all kinds, *i.e.* detritus; and above-ground NPP) enables an estimate to be made of the amount of C internally translocated below ground. A mass balance of the component fluxes below ground, comprising the inputs from above (*i.e.* the detritus and translocate), root and heterotrophic respiration, fine-root turnover, and mycorrhizal and root system NPP, enables an approximate estimate to be made of net changes in the pool of soil C.

In principle, summation of the component fluxes over a year should add up to the independently measured annual NEE. However, there are appreciable errors in measuring the below-ground components in particular, so that close agreement between measurements or estimates of the components of the budget is not as yet to be expected.

Exchanges of Volatile Organic Compounds and Methane

The net exchange of C between forests and the atmosphere also includes the emission of bio-genic volatile organic C compounds (VOC) such as isoprene and terpenes from the tree canopy. Simple algorithms have been developed to calculate emissions of individual compounds from individual tree species based mainly on PAR and leaf temperature. The emissions of VOC increase with temperature and PPFD and it has been argued that the emissions of certain VOC compounds, notably isoprene, reduce the physiological effects of water stress. The absolute magnitude of C exchange represents a small fraction of the total. Emissions of terpenes from Norway spruce in Denmark amount to no more than 0.3% of the annual C assimilated, but short-term losses from the Mediterranean species *Quercus ilex* may exceed 6% at high temperatures Globally emissions are small, around 0.1% of annual carbon emissions from vegetation.

Undrained peatlands in high latitudes have accumulated appreciable amounts of C from the atmosphere since the retreat of the ice and are currently significant annual CO_2 sinks (0.2-0.5Mgha^{-1} of C) but are also small annual sources of methane (CH_4) (0.03-0.3 Mgha^{-1} of C). By contrast, drained peatlands, whether intended for agriculture or afforestation, release C as CO_2 because of accelerated decomposition of the aerobic peat but no longer release CH_4 in significant amounts. Peatlands drained for agriculture continue to be a sustained C source while the peat remains. Depending on species of tree and time in the rotation, peatlands drained for afforestation may continue to be a source of C in spite of forest biomass growth. Thus drainage and afforestation of peatlands reduces CH_4 emissions and provides a larger CO_2 sink than the peatland plant community. However, the net exchange of C by peatlands is an upward flux for some years, until the C fixed by the canopy exceeds the losses of C from the oxidizing peat with the lower water table. The primary interest in forest-atmosphere C fluxes is the contribution to the global atmosphere C budgets and the consequent radiative forcing of climate. As CH_4 on a per molecule basis in radiative equivalents has a global warming potential 30 times that of CO_2, it is not simply the mass fraction that should be used to identify priorities. The change in the net direction of CH_4 flux from emissions to deposition needs to be included in C budgets and the overall net radiative forcing calculated.

Future Carbon Sequestration Potential

Across Europe forests are now growing faster than previously and similarly in the tropics. We currently seem to be in a period of especially strong C uptake by forests, although it has been suggested that this will not continue indefinitely, and indeed there is concern that the rate will diminish significantly as the 21st century proceeds.

It has frequently been suggested that the present rates of CO_2 uptake and growth are the combined result of the recent global increases in atmospheric CO_2 concentration , N deposition and temperature. Quantitative evaluation of the relative impacts of all three of these factors is a focus of current research.

CO_2-stimulated Carbon Sequestration

It is difficult to measure the impact of the slow progressive rise in atmospheric CO_2 concentration on the C sequestration capacity of stands of trees but some relevant indications can be obtained from recent experimental programmes to evaluate the impacts of elevated atmospheric CO_2 concentration on the growth, development and physiology of young trees.

Doubling atmospheric CO_2 concentration speeds up development, photosynthesis and growth by up to 60% in a wide range of species of young trees grown rooted in the ground in the field over periods of up to 5 years in both North America and Europe. Moderate lack of nutrients and of water reduce growth but have little adverse effect on the *relative* impact of the increase in CO_2 concentration. To all intents and purposes the impact of the increase in CO_2 concentration is that trees get bigger quicker but otherwise remain similar in most respects to trees of the same size in the current conditions. Meta-analysis of the results from many experiments shows that the key physiological parameters defining capacity for photosyn-thetic CO_2 uptake by trees, V_{cmax} and J_{max}, are downregulated on average by no more than 12% and stomatal conductance by 15% as a result of growth in elevated CO_2 concentration.

A key question, as yet unresolved, is to what extent will the responses of key parameters to increased CO_2 concentration change when a stand of young trees reaches canopy closure. This is not readily solvable other than by long-term, free air elevated CO_2 concentration exposure experiments (FACE), starting with seedlings or clonal propagules. So far, the few FACE experiments in progress in forests have been imposed on existing stands (*e.g.* Ellsworth 1999), which are unlikely to acclimate fully over, say, a 5-year period of exposure. Thus major uncertainties remain as to the likely long-term effects of the rise in atmospheric CO_2 concentration on C sequestration and the contribution of CO_2 fertilization to the global terrestrial C sink.

Temperature-stimulated Carbon Return to the Atmosphere

Photosynthesis is particularly dependent on solar radiation, CO_2 concentration and N availability and thus is likely to be reduced by increase in cloudiness but increased by elevations of the global atmospheric CO_2 concentration and by N deposition. All respiratory processes are very sensitive to temperature, as is the multiplication of the populations of respiring organisms, particularly the fine roots and heterotrophs in the soil. Thus soil respiration increases strongly in boreal and temperate forests in relation to the rise in soil temperature as the spring progresses and consequently is likely to be very sensitive to the increase in global temperature. Increase in soil temperature leads to enhanced mineralization of soil organic matter and release of N, which feed back to stimulate photosynthesis, tree growth and transfer of CO_2 from atmosphere to forest. The balance between these conflicting effects of rising temperature on the net exchange of CO_2 (enhanced respiration and enhanced photosynthesis) is only now being resolved experimentally in the short term, although models indicate a potentially complex time-course in the long term.

Carbon Sink Saturation

It has been proposed that on time-scales of decades to centuries saturation of C sinks will occur and that existing C sinks will become significant C sources. This effect, it is suggested, arises from a difference in phase and functional relationship between the responses to global climate change of the principal gain and loss terms in NEE. The CO_2 stimulation of photosynthesis increases with increasing atmospheric CO_2 concentration (the CO_2-fertilization effect) but does so at a diminishing rate. On the other hand, heterotrophic respiration (decomposition of litter and soil C, resulting in emission of CO_2 to the atmosphere) increases somewhat later in time as detritus production builds up and increases exponentially with increasing temperature.

Because so much C is stored in forest soils, and because this C is potentially vulnerable to climate change, it is essential to take full account of the C stocks in soil. It has been postulated on the basis of models that changes in climate, particularly the increase in temperature, will lead to higher rates of heterotrophic respiration and oxidation of the C stock in the soils, it is presumed without acclimation. Thus, as both atmospheric CO_2 concentration and temperature rise, the rate of heterotrophic respiration may be expected to increase relative to the rate of photosynthesis, and the overall capacity of the terrestrial biosphere to take up C from the atmosphere to diminish. However there are a number of possible feedbacks not included in these model analyses. These include the effects of rising CO_2 and temperature on:

1 litter quantity, which may increase as a result of any short-term CO_2-fertilization effect and may lead to enhanced nutrient availability and C storage in soils;
2 litter quality (*i.e.* the C:N ratio), which may decrease and retard decomposition and nutrient release;
3 allocation of C below ground, which may well increase and lead to increased production of roots, mycorrhizae and exudates;
4 increased nutrient availability and temperature leading to higher photosynthetic rates and NPP; and
5 longer photosynthetic and growing seasons.

These feedbacks raise major uncertainties, currently being addressed by soil-warming experiments and modelling. The results of recent modelling analyses that have taken these feedbacks into account suggest that the dominant *short-term* constraint on the CO_2-fertilization effect is an increase in litter quantity. On the other hand, *long-term* increases in C sequestration are possible as atmospheric CO_2 concentration increases, if the net N input also increases (either through fixation or deposition) or if the N : C composition of newly formed litter substantially decreases. If below-ground C allocation increases, as observed in many experiments, and leads to increased production of roots, mycorrhizae and exudates, with consequent increase in N fixation, a large long-term increase in NPP and C sequestration is likely. The simulated *long-term* effect of rising temperature is a large increase in NPP, and hence litter production, as a result of enhanced N mineralization and consequent increased photosynthesis, whereas *short-term* effects of temperature on photosynthesis and respiration are relatively unimportant.

These simulations highlight the importance of litter production on the one hand and turnover processes in the soil on the other, the former being strongly constrained by N availability and air temperature, the latter by soil microbiological processes and soil temperature. One cannot escape the conclusion that more emphasis needs to be placed on understanding better the processes in the soil, and the link between soil and atmosphere through the medium of the trees.

Indications from soil-warming experiments are that after an early initial flush of CO_2, the CO_2 efflux is similar in the heated and unheated control plots; furthermore, the feedbacks are actually increasing the efficiency of the 'forest pump' through the release of N and may override the initial losses of C. The effect of climate change on the efficiency of the pump in transferring C from the atmosphere to the soil via enhanced nutrient availability, photosynthesis and detritus production needs to be evaluated in addition to the effect of climate change on the return of C from the soil to the atmosphere, in order to obtain a complete picture of the net change in C sequestration by forests in a warmer climate.

A further key question is whether the current annual rate of N deposition (= $20 kgha-^{1}$) adequately balances the current annual rise in atmospheric CO_2 concentration (= $1.5 mmolmol^{-1}$) to sustain the NEE over the next 50 years. As things are at the present, the possibility of C sinks becoming sources in the mid twenty-first century remains highly speculative.

EVAPORATION AND TRANSPIRATION

Evaporation and transpiration of water vapour from forests have major influences on forest production, local water supplies and regional climates. Depending on the precipitation climate, significant amounts of water are returned to the atmosphere by evaporation after interception by the overstorey and understorey canopies without ever reaching the soil, the so-called interception loss. A proportion of the precipitation reaches the ground by through-fall and stemflow and infiltrates the soil. Some of this water is taken up by roots of the trees and understorey vegetation, and having traversed from roots to leaves is returned to the atmosphere in transpiration, defined as evaporation of water from within the leaves and its subsequent diffusion and turbulent transfer from the sites of evaporation to the air above. Total evaporation and transpiration rates from forests may reach 7 mm daily, although annual totals mostly average 1-3 mm daily.

Driving Variables and Constraints

The driving variables for both the evaporation of intercepted water and for transpiration are the net radiation and sensible heat absorbed by the leaves, both of which raise leaf temperature and power the phase change in evaporation from liquid water on leaf surfaces or within leaves to water vapour, together with the water vapour mole fraction or vapour pressure deficit at the sink for water vapour in the atmosphere outwith the canopy. The process is constrained by the catena of conductances or resistances along the pathway from the sources of water vapour on the surfaces of wet leaves or from within dry leaves to the sink in the atmosphere. This catena of conductances, together with the difference in water vapour mole fraction between the sources of water vapour on the surfaces of, or within, the canopy leaves and the sink in the atmosphere, determines the tendency for molecules of water vapour to escape from the canopy to the atmosphere. Surface conductances (or resistances) strongly determine the exchange of water and other scalars between vegetation and the atmosphere at leaf, stand, forest and regional scales.

Evaporation of Intercepted Water

Although models using a statistical description of raindrop size and frequency may give a somewhat better predictive representation of the interception process, the evaporation of water from wet canopies (E_i) has generally been represented using the Penman equation in the Rutter model, *i.e.*

E_i = [e/(e + 1)](*A*/A) + [1(e + *1)]{g_aD/p)*

where e is a coefficient for change in the sensible and latent heat contents of air with respect to temperature, *X* the molar latent heat of vaporization of water, and *D* and *P* the ambient water vapour pressure saturation deficit and atmospheric pressure of the air at the reference location above the canopy. Both the available energy and saturation deficit of the atmosphere are usually small (< 100Wm^{-2} and < 0.1 kPa, respectively) during and immediately after rainfall events, when evaporation of intercepted water occurs. None the less, evaporation rates are high, up to 0.6 mm h^{-1} (400 W m^{-2}, 10 mmol m^{-2} s^{-1}) for short periods as measured by eddy covariance or inversion of the Penman equation. This is a result of the high aerodynamic conductance to sensible heat transfer towards the forest surface, and to water vapour transfer away from the source to the sink in the atmosphere above, and because the saturation deficit, although small, is maintained by advection of drier air from the landscape surrounding the area of the rainstorm.

The water vapour pressure at the canopy surface may fall to the saturation vapour pressure at the wet-bulb temperature and then the wet canopy behaves like a large psychrometer: the rate of evaporation is balanced by the input of sensible heat and thus given solely by the second term on the right-hand side of Eqn 10.5, *i.e.* the net radiation term can be neglected, a useful convenience when deriving g_a from inversion.

The proportion of gross precipitation that is intercepted and directly evaporated back to the atmosphere depends strongly on the frequency and intensity of the rainfall. In the tropics, where much of the rainfall comes in relatively short intense storms, interception loss may amount to no more than 11% of the gross precipitation, although some higher values have been reported.

At the other extreme, in maritime climates where the rainfall comes in gentle showers over long periods, so that the canopy is wet for much of the time, interception loss can amount to 40% of the measured precipitation.

Furthermore, much of the fog and, on hills and mountains, the cloud water intercepted by the foliage of trees is not measured by conventional rain gauges and is readily evaporated back to the atmosphere, frequently leaving a nasty deposit on leaf surfaces. In 'cloud forest' however, fog days may add significantly to the precipitation, although not measured as such, and consequently is known as 'occult' precipitation.

The evaporation of intercepted water is influenced by tree spacing and logging. One might expect the amount of water evaporated to decline in proportion to the reduction in number of trees because of the reduction in surface area of wet foliage. However, in a Sitka spruce spacing experiment this was found to be compensated to a significant degree by the increase in g_a on a per tree basis associated with the increase in distance between the trees and similar results were obtained in a comparison between logged and unlogged areas of tropical forest.

Valente *et al.* (1997) added an additional coefficient to the Rutter model to take account of the wide open spaces in the sparse canopies of Mediterranean forests and Gash *et al.* (1999) have shown, by comparison with eddy covariance data, that evaporation is extremely well represented by their sparse-canopy Rutter model in combination with the *momentum* aerodynamic conductance.

Evaporation of Transpired Water

A forest comprises a system of parallel conduits through which water is passively transferred from soil to atmosphere as the result of evaporation of water from the leaves and the consequent development of a gradient in water potential. Recently, attention has become strongly focused on the capacity of the liquid water conducting system and its vulnerability to cavitation and entry of air into the vessels and tracheids.

However, it is relevant to re-emphasize the point made long ago that by far and away the largest drop in water potential (or equivalent vapour pressure) in the soil-plant-atmosphere continuum (SPAC) occurs across the stomatal pores, and consequently this is the one location in the SPAC where the flow of water from soil to atmosphere can effectively be regulated.

Thus the conductances of the stomata in a dry canopy provide the fine-scale, variable control of the transpiration and latent heat fluxes. While the mechanism of stomatal action is still somewhat of a 'black box', one hypothesis that has been incorporated into some models has it that stomatal conductance adjusts to minimize the risk that the conducting system is impaired through cavitation and air entry.

The driving forces and constraints on transpiration are combined in the now very familiar Penman-Monteith equation, which provides the basis for the representation of transpiration in most canopy models. While based on sound physics at the scales of the pore and the leaf, use of this equation at canopy scale does involve some assumptions and approximations because of the different spatial distribution of the sources and sinks for heat and water vapour in a canopy compared with a single leaf.

Coupling to the Atmosphere

While the temperature and humidity of the air may in some circumstances be regarded as independent driving variables, vegetation influences its own local environment to varying degrees, depending particularly on its structure. At onc extreme, feedback from the exchanges of heat and water vapour in extensive areas of crops, for example, determines the temperature and humidity within the vegetation and in the adjacent surface layer. Equilibrium profiles of temperature and humidity develop in relation to the fluxes of heat and water vapour that are driven by the absorbed net radiation. Consider the consequences of a change in conditions that lead to an increase in transpiration, such as an increase in stomatal conductance or saturation deficit. There will be a transient increase in transpiration, but if the additional molecules of water vapour are not able to escape from the vicinity of the foliage, transpiration will subsequently be damped down and will return to its original rate, as determined by the net radiation. From an analysis of transpiration datasets obtained from extensive areas of field crops, Priestley and Taylor (1972) arrived at the following model for equilibrium transpiration:

Eeq = *[£/(£+ 1)\cc(A/X)*

where a is a coefficient, in principle unity but found by Priestley and Taylor to have an empirical value of about 1.3 for field crops, that essentially takes into account effects of entrainment and advection when this equation is used alone. More recently, this model for transpiration has been shown to represent transpiration by a range of field crops reasonably well but was found to be completely inadequate for plantations of trees, for which wholly improbable values of *a* of about 11 were required.

At the other extreme, consider a stand of tall trees with an irregular, 'rough' canopy surface. Partly because the foliage of tall trees is in a higher wind-speed regime but also because of the 'rough' surface, air movement and turbulence are much greater among the foliage, with the result that the ambient environmental conditions are imposed close to the leaf surfaces. Consequently, when an initial increase in transpiration occurs because of an increase in stomatal conductance or ambient saturation deficit, the additional molecules of water vapour have little difficulty in escaping so that the increase in transpiration is maintained at a new higher rate that is largely independent of the net radiation input. At this limit, leaf and air temperatures are equal and the ambient saturation deficit is imposed right up to the leaf surfaces so that transpiration is given by

$E_{imp} = gcD/(P)$

McNaughton and Jarvis (1983) and Jarvis and McNaughton (1986) redeveloped the Penman-Monteith equation to give expression to these two limit situations and in the process derived a decoupling coefficient, *Q*., that defines the extent to which actual transpiration approaches one or other of the two limits, *i.e.*

E = *GE*,,.

. -(1-J2)E_{im}P

where Q = (e+ 1)/(e+ 1+g_a/g_c)

Q has a range of zero to 1.0, approaching zero as g_a increases relative to g_c and approaching 1.0 as g_c increases relative to g_a. Since g_a is a function of wind speed and turbulence and g_c a function of water stress, the ratio g_a/g_c is large for tall stands of trees in windy environments, particularly when saturation deficits are large and soil water stress intervenes, and is small for well-watered short vegetation in less windy conditions. The ratio also varies somewhat with stand and canopy structural features such as tree spacing and the spatial distribution of leaf area density, both of which affect g_a, and with leaf area, structure, anatomy and physiology, all of which affect g_c. Estimates of *Q* are in the range 0.1-0.3 for stands of conifers and 0.3-0.6 for stands of broadleaved trees, the lower values of each range being in windier maritime climates, the higher values in less windy, more continental climates. These low values of *Q* show that E_{im} is the much more important component of transpiration in forests than is E_{eq} and thus that saturation deficit is a more important driving variable than net radiation, especially in conifers, and indeed that stomatal conductance is a more important constraint than is the availability of absorbed net radiation. Furthermore, *E* is necessarily limited by the availability of net radiation whereas E_{im} is open-ended in response to saturation deficit, unless or until stomatal closure or leaf fall intervenes.

Dynamics of Transpiration

The daily dynamics of forest transpiration are driven in particular by the saturation deficit, and to a lesser extent by the net radiation, and follow very closely changes in surface conductance, which are moderated especially by the fluctuating incident PPFD and saturation deficit of the ambient atmosphere and by the availability of soil water.

The seasonal dynamics are also strongly constrained by the surface conductance, but for rather different reasons in different localities. In the boreal forests, the stomata remain closed during the winter while the soil is frozen so that it is the timing of the thaw in the spring and the freeze-up in the autumn that determines the period over which the surface conductance can respond to the environmental driving variables.

In deciduous temperate forests, the phenology of leafing and senescence of the foliage determines the period of transpiration, and in Mediterranean climates the seasonal dynamics are determined by lack of water. In humid tropical forests, the seasonality is much less evident, although even here periods of water stress occur and the canopy leaf area also fluctuates somewhat for other reasons, thereby affecting the surface conductance.

Role of the Understorey

In widely spaced open stands the evaporating surface area, and hence the canopy conductance, of the trees is less than in denser stands and it might be thought that the amounts of water returned to the atmosphere in transpiration would be correspondingly less. However, Roberts (1983) compiled data on annual transpiration from a number of investigations in stands of different kinds and stocking across Europe, including some of his own, and found that the annual amounts of transpiration were very similar among the stands, with an overall mean of 333+35 (s.d.) mm (n= 12). He drew the conclusion that strong development of the understorey in widely spaced stands compensated for the reduction in the overstorey as far as transpiration is concerned. Is this likely? A priori one might assume that there would be little transpiration from the understorey, firstly because it is shaded from above and consequently has less net absorbed radiation, and secondly because it would be much less well coupled to the atmosphere above the stand. In the latter case, transpiration would tend more towards the equilibrium rate and would thus be small because of the small input of net radiation.

However, measurements of vertical profiles of air temperature, water vapour pressure, CO_2 and other trace gas concentrations, particularly in coniferous stands, show only very small differences with height at any one time and measurements of these variables in the trunk space differ little on average from those above the canopy. The explanation for this lies in the frequent penetrating downdrafts of air from above every few minutes. In other words, the understories of coniferous stands are surprisingly well coupled to the atmosphere above. The combination of good coupling with the comparatively high stomatal conductances in understorey foliage seems to be sufficient to account for the compensation noted by Roberts (1983). The degree of coupling of the understorey to the atmosphere above is somewhat less in stands of broadleaved trees, possibly because of less deep canopies with higher leaf area density, and is very much less in multistoried, humid, tropical forest stands where there are bands of high leaf area density in the upper canopy.

Bowen Ratio

A functional descriptor of the partitioning of energy between warming or cooling the forest and evaporating water, and varies widely depending on forest type and climate. Analysis of *[3* in terms of the likely variables shows that these observations are explicable in terms of the so-called 'isothermal' or 'climatological' resistance of Monteith (1965) and Stewart and Thom (1973). Combining this resistance with the Penman-Monteith equation leads to the following expression for the Bowen ratio

$$p+1 = (e+1 + gag_c)/(e + gag_i)$$

where the climatological conductance, g_i, is given by:

$$g_i=(AfD)-(P/X)$$

It is apparent that p is approximately proportional to the climatological conductance, g_i, essentially the ratio A/D, and inversely proportional to the stomatal conductance, becoming larger in a predictable manner as D declines relative to A, or as g_c changes with vegetation type and structure and as stomata close. In the case of different climates cited earlier, it is the differences in g_i that are important; in the case of different forest types, it is the differences in g_c among the stands that account for the different values of p.

Equation 10.10 can be rearranged to provide a simple direct means of deriving the canopy conductance from a knowledge of p, g_a and g_i as follows:

$$g_c = 1[(\#-1)/g_a+(/J+1)/g_i]$$

This is not only a useful means of derivation of g_c but also provides a practical description of the interrelationships among energy partitioning, the pathway properties and the driving variables of available energy and saturation deficit. In essence this expression encapsulates the most significant functional attributes of forest canopies with respect to the exchanges of heat and water vapour with the atmosphere.

Some Conclusions

There are some paradoxes evident here with respect to water use by forests. Canopy aerodynamic conductances are generally high, especially so in comparison with crops, grassland and heath-land, decreasing somewhat from coniferous boreal forest, through mixed broadleaved temperate forest to multistorey, humid, tropical forest. Conversely, canopy surface conductances are generally low in coniferous boreal forest, close to zero during the period of frozen soils in winter; somewhat higher in broadleaved temperate forest, at least during the foliated period; and possibly somewhat higher still in humid tropical forest, the year round. In general, forest canopies, particularly coniferous forest canopies with large g_a/g_c, are extremely well coupled to the atmosphere. High g_a leads to high rates of evaporation of intercepted water from canopy surfaces; low g_c leads to low rates of transfer of transpired water vapour from canopies to the atmosphere. Thus in climates where a large proportion of the precipitation occurs during long periods of low-intensity rainfall, such as in temperate maritime regions, evaporation of intercepted rainfall, particularly from coniferous forest, can amount to a very large direct return of water to the atmosphere. On the other hand, the low g_c of temperate and boreal coniferous forest results in low transpiration rates from dry canopies. Thus the local forest water budget depends on the balance between the periods of time that the canopy is wet or dry and thus on the precipitation regime in a particular area.

Because of low g_c, forest canopies, even when not superficially wet, present a particularly dry surface to the atmosphere. High g_a/g_c also leads to relatively high rates of sensible heat transfer and to large Bowen ratios which drive regional climate. However, the magnitude of this effect is moderated by another aspect of local climate, the relative size of g_i, the climatological conductance. At identical g_c, p increases as g_i increases and transpiration is inhibited by small saturation deficits.

We conclude this book on evaporation and transpiration, with its emphasis on Q and p, by suggesting that these two coefficients are useful summarizing descriptors of the functional properties of forest stands and could well form the basis for a classification of forest stands into functional types.

EXCHANGE OF SCALARS

Exchanges of scalars (water vapour, trace gases, particles) between forests and the atmosphere can be expressed as a flux (F) driven by a difference in 'concentration' (AX) and constrained by a conductance (g) or resistance (r), *i.e.*

$$F = g\text{-}AX = AX/r$$

Depending on the scalar and whether the net flux is towards or away from the canopy, the flux is constrained by a catena of conductances or resistances along the pathway between sources or sinks on the surfaces of leaves, twigs and branches, or within leaves, and the source or sink in the atmosphere, notably the conductances of the stomata and stomatal antechambers, the conductance of the leaf cuticle and other tree surfaces, the conductances across the leaf boundary layers, and the aerodynamic conductances through the canopy and across the canopy boundary layer to a reference location above.

While the application of conductance and resistance approaches are the norm for current applications, it is necessary in important cases to recognize that fluxes are bidirectional and that multiple sources and sinks within forest canopies increasingly require more complex treatments to describe the interaction of biological production, uptake and chemical processing with the turbulent exchange processes above and within canopies. At stand scale, canopy stomatal and boundary layer conductances are derived as the product of the leaf conductances and the leaf area index of leaves, appropriately weighted according to the scalar. The scaled-up leaf stomatal conductance is commonly referred to as the canopy surface conductance, or simply the canopy conductance.

Historically, resistances (the inverse of conductances) were first used but recently conductances have become more widespread. The conductance terminology is now commonly used for transfer of CO_2 and, to a lesser extent, water vapour, whereas the preferred convention for quantifying the affinity of vegetation surfaces for uptake of *reactive gases* remains the canopy or surface resistance , because the full range of surface affinities must be included and this would result in infinite surface conductances. The resistance approach is therefore not only conventional but also preferable on theoretical grounds.

In the cases of both HNO_3 and HCl, for example, the surface resistance is negligible and tree surfaces generally are considered to be 'perfect sinks' for these gases. There are few measurements of HNO_3 (and especially HCl) canopy fluxes, although the work so far indicates zero canopy resistance for these gases.

The combination of large atmospheric conductance (g_a) (or small atmospheric resistance, r_a) and zero canopy resistance (r_c) leads to significant inputs of atmospheric nitrogen as HNO_3 to forests even when ambient concentrations are very small and, conversely, to large losses of intercepted precipitation from wet canopies although atmospheric saturation deficits are then small.

More problematical, at least two sets of units have been, and continue to be, widely used by different sectors of the community, depending largely on whether fluxes are required in units of mass (kg), volume (m^3) or amount (mol) and whether 'concentration' differences are expressed as mixing ratios, volume fractions, mole fractions, partial pressures, etc. Conveniently, where concentrations are dimensionless, as in the case of mole fraction for example, the conductance has the same dimensions as the flux and the dimensions of the resistance are the inverse. Where the 'flux' is expressed as a volume flux density, *i.e.* $m^3\ m^{-2}\ s^{-1}$, the conductance has the units of $m\ s^{-1}$ and the resistance the somewhat curious units of sm^{-1}!

Aerodynamic Conductance/resistance

The aerodynamic conductance/resistance should be formulated to give an accurate representation of the transfer of the appropriate scalar of interest (water vapour, CO_2 and other trace gases) from the source or sink in the canopy, across the atmospheric surface layer to the sink or source in the mixed layer.

Aerodynamic conductances of canopies (g_a) have most frequently been estimated using the transfer of momentum to canopies or the transfer of water vapour away from wet canopies as tracers.

The transfer of momentum is driven by pressure forces and all canopy elements provide the sink, whereas the transfer of water vapour and trace gases is by molecular or turbulent diffusion and the sources and sinks in the canopy are usually only the leaves, depending on the scalar. The difference between the two can be accounted for but since the aerodynamic resistance of forest canopies is generally at least one order of magnitude smaller than the surface resistance because of the strong turbulence in and near forest canopies, little is gained by this and either may be used.

Canopy aerodynamic conductance increases in proportion to wind speed. Analysis of wind profiles showed g_a for momentum to increase from 0.08 m s^{-1} (3.2molm^{-2}s^{-1}) for wind speed of 1 ms^{-1} at canopy height to 0.5ms^{-1} (20molm–^{2}s^{-1}) for wind speed of 6 m s^{-1} above a Sitka spruce stand. From a review of wind profile data, the following general relation was deduced for coniferous forest stands :

$g_a = 0.1u(\text{A})\text{m s}^{-1}$, *i.e.* $g_a = 4u(\text{A})\text{molm}^{\sim 2}\ \text{s}^{-1}$

where $u(h)$ is wind speed in ms^{-1} at canopy height.

In a Sitka spruce spacing experiment g_a for water vapour was found to increase asymptotically in relation to number of trees, and hence leaf area, per hectare, whereas the conductance on a per tree crown basis increased linearly with distance between the trees.

This can be attributed to enhanced ventilation of the trees and increases in a number of turbulence parameters (Green 1990). Typical values of g_a for water vapour were in the range 3-9 mol m^{-2} s^{-1}. In a comparison between selectively logged and unlogged plots in tropical forest, Asdak *et al.* (1998a,b) found a comparable situation: the boundary layer conductance, on a per tree basis, of the trees remaining on the logged plot was higher than that of the trees on the comparable unlogged plot.

It seems likely that g_a will also be influenced by vertical distribution of leaf area density in canopies: a high leaf area density in the upper layers of a multistoried tropical forest canopy inhibits penetration of downdrafts and effectively decouples the lower layers from the air above. However, effects of crown and canopy structure on g_a have not yet been systematically investigated.

Surface Conductance/resistance

Surface conductances of canopies (g_c) have frequently been estimated using water vapour transfer from dry canopies as a tracer and scaled for other trace gases such as CO_2 by the ratio of their molecular diffusivities. For a range of deposition fluxes, using a resistance analogy allows the aerodynamic resistance to be subtracted from the total transfer resistance to give the surface resistance. The absence of a residual term implies very efficient exchange and large deposition rates (velocities). However, more commonly a residual or surface resistance is present which, in the case of reactive trace gases, may be considered as a measure of the chemical affinity of the absorbing surface for the trace gas in question. In general, highly reactive gases such as HCl or HNO_3 are characterized by very small or zero surface resistances.

Surface conductances of dry canopies for water vapour (g_c) lie in the range 0.02-2molm^{-2}s^{-1}. Despite high leaf area index in some species, very low maximum values of g_c occur in conifers, partly because the stomatal antechambers are filled with wax rods and tubules. Higher values of g_c occur in broadleaved trees, particularly in multistorey tropical forest. In addition to the area of leaves in a canopy, g_c is sensitive to the several environmental and physiological variables that influence stomatal aperture. Stomatal conductance increases with PPFD, has a temperature optimum, and decreases with atmospheric water vapour saturation deficit and CO_2 concentration, and with shortage of soil water. Canopy surface conductances are usually at least an order of magnitude smaller than canopy aerodynamic conductances so that fluxes from forest canopies are generally effectively constrained by g_c. In coniferous forests in particular, atmospheric conditions at the leaf and canopy surfaces are very similar to the conditions overhead so that feedbacks affecting the local atmosphere play little or no role in constraining transpiration and exchanges of non-reactive trace gases, which are highly sensitive to changes in g_c.

There are several alternative models of stomatal action in relation to environmental variables and leaf properties and these have been incorporated into large-scale soil-vegetation atmosphere transfer schemes (SVATS). In due course dynamic models incorporating surface chemistry will be able to simulate surface reactive processes within SVATS, and the first steps in this process have been made. However, for most forests and trace gases, current understanding limits the scope for such modelling to a very few specific cases.

CARBON BALANCE ANALYSES

The concurrent effects of increasing atmospheric CO_2 concentration, climate variability, and cropland establishment and abandonment on terrestrial carbon storage between 1920 and 1992 were assessed using a standard simulation protocol with four process-based terrestrial biosphere models. Over the long-term(1920–1992), the simulations yielded a time history of terrestrial uptake that is consistent (within the uncertainty) with a long-term analysis based on ice core and atmospheric CO_2 data. With increased concerns about how forests may respond to changes in atmospheric chemistry and climate, there has been a major effort to gain a better understanding of the processes controlling tree growth. This effort has been accentuated by observations that historical measures of production no longer apply, possibly as a result of changes in genetic composition of trees, application of fertilizers, control of competing vegetation and alterations in atmospheric chemistry (CO_2, ozone, acid rain, etc.). Developing a more integrated picture of how processes such as photosynthesis, respiration and allocation combine to affect the structure and function of the autotrophic component of forest ecosystems is critical to predicting forest growth under changing global conditions. Although net primary production represents all carbon products produced annually from photosynthesis after subtracting losses from respiration, the components of respiration and growth vary significantly depending on the environment, composition of tissue, and allocation of resources above and below ground.

Figure: Analysis in Carbon Balance

Canopy Photosynthesis

Photosynthesis is among the best understood of all physiological processes. Historically, most research on photosynthesis has been done under laboratory conditions on individual trees. A series of basic equations have been developed for individual leaves that describe the capture of light energy, the diffusion of CO_2 and the enzymatic transformations involved in the photosynthetic process. To apply these basic equations to whole canopies involves an analysis of how photosynthetic capacity and environment vary spatially. Sophisticated models have been constructed to predict the quantity of direct and diffuse radiation absorbed within a forest canopy.

Although such models are useful for predicting the maximum rates of photosynthesis, they are not able to predict rates under varying environments. To extend estimates of photosynthesis to whole canopies two important simplifications are generally made: (i) the canopy is considered to represent a single large leaf with varying ability to absorb radiation; and (ii) the relation between photosynthesis and light absorption is extended in time to achieve a more linear response than occurs instantaneously. Further simplifications are possible by combining the effects of subfreezing temperatures, drought, vapour pressure deficits and nutritional limitations on the conversion of absorbed radiation into photosynthetic products into a general equation as suggested by Landsberg (1986).

Autotrophic Respiration

A large fraction of gross photosynthesis is expended through autotrophic respiration in the construction and maintenance of plant tissue. Physiologists generally separate respiration into three major components: photorespiration *(R)*, construction respiration *(R_c)* and maintenance respiration *(R_m)*.

Photorespiration

In light, photorespiration occurs in the process of generating the substrate for the enzyme ribulose bisphosphate carboxylase-oxygenase (Rubisco), essential for the primary fixation of CO_2. Under conditions where the CO_2 concentration within the leaf is depleted relative to the concentration of O_2, the enzyme releases CO_2 rather than fixing it into carbohydrates. In general, higher temperatures stimulate photorespiration to the extent that simple estimates of gross photosynthesis may be overestimated by as much as 10%. Conversely, rising CO_2 concentrations in the atmosphere tend to moderate the effects of rising temperature on photorespiration and shift the optimum temperature for photosynthesis upward. At temperatures above 40°C, gross photosynthesis decreases abruptly because of changes in chloroplast and enzyme activity.

Construction respiration

The synthesis of plant tissue requires the metabolism of more resources than found in the final product. Rates of construction respiration for various tissues differ, depending on their composition. From an analysis of biochemical pathways Penning de Vries (1975) estimated that the production of 1 g of lipid would require 3.02g of glucose, whereas 1 g of lignin, protein or sugar polymer might require 1.90, 2.35 and 1.18 g of glucose respectively. More recently, empirical relations have been derived to estimate the total cost of construction based on a correlation with the heat of combustion, ash, and organic nitrogen content of tissue and its biochemical composition.

In general, 1.25-1.35 g of carbon are required to produce tissue containing 1 g of carbon. Although the amount of tissue produced may be a function of temperature, the construction cost per gram is fixed. This fact makes it possible to assess construction respiration at annual time steps from forest inventory data using allometric relations between measured increases in stem diameter and the production of foliage, branches, stemwood and large-diameter roots.

Maintenance Respiration

Maintenance respiration *(R_m)*, the basal rate of metabolism, includes the energy expended on ion uptake and transfer within plants. Repair of injured tissue may greatly increase respiration above the basal rate. Because trees accumulate a large amount of conducting and storage tissue as they age, the observed decrease in relative growth rates associated with age has often been assumed to reflect increasing maintenance costs. However, most of the tissue in trees is wood and, of this, only the sapwood contains any living cells (5-30%). It is unlikely that the metabolism of this extra sapwood is enough to limit growth. Enzymatic activity associated with protein turnover can be predicted as a function of nitrogen content, the fraction of living cells in foliage, branches, stems and roots, and temperature of the tissue. Maintenance respiration generally increases exponentially with temperature within the range of biological activity, as described by the formula

$$R_m(T) = R_0 Q_{10}$$

where R_0 is the basal respiration rate at $T = 0°C$ (or other reference temperature) and Q_{10} is the respiration quotient, which represents the change in the rate of respiration for a 10°C change in temperature (T). The Q_{10} is usually about 2.0-2.3. Under field conditions, where daily temperature variation is large, the non-linearity in the response of maintenance respiration can be important because the respiration rate will depend on the pattern of temperature variation even in sites with the same mean temperature. This has been modelled by fitting a sine function to minimum-maximum temperature differences and integrating the response not only daily but annually. From analyses made in pine forests growing at mean annual temperatures from 5 to >20°C, Ryan *et al.* (1995) concluded that the annual above-ground woody tissue maintenance respiration represents 5-15% of gross photosynthesis and that other factors must therefore be responsible for commonly observed reductions of more than 50% in above-ground production as forests age.

Net Primary Production

Annual net primary production (NPP) is conceptually easy to define as the residual carbon products remaining after autrotrophic respiration has been subtracted from gross photosynthesis. In practice, it is difficult to obtain accurate estimates of seasonal changes in foliage and fine-root mass, variation in metabolic reserves, and precise measurements of tissue temperatures throughout a forest, thus making estimates of NPP difficult. None the less, carbon balance estimates that take into account construction and maintenance respiration along with estimates of annual growth come fairly close to independent estimates of gross primary production (GPP), modelled or derived from microme-teorological measurements.

If one assumes that half the carbon allocated below ground goes into the production of roots while the rest is expended in construction and maintenance respiration, an estimate of total NPP can be made from datasets. When this was done for 12 temperate forests, the ratio NPP: GPP approached a constant, 0.47 + 0.04. Similar analyses in boreal forests indicate that NPP : GPP = 0.3.

This lower ratio in the boreal forest zone is attributed to delayed recovery in photosynthetic capacity of evergreen canopies in spring when solar radiation peaks. Similar constant relationships between the ratio of autotrophic respiration and gross photosynthesis have been reported in growth-room experiments where pine and eucalypt seedlings have been raised under a range of temperatures from 15 to 25°C. It is important to emphasize that the fraction of carbon allocated to wood production can vary by more than threefold as a result of differences in the way that carbon resources are allocated to growth.

Allocation of Carbon Resources

Although the overall NPP : GPP ratio derived from carbon balance analyses may approach a constant, the proportion of carbon allocated to roots increases with climatic limitations on gross photosynthesis. The availability of nutrients alone can alter above- and below-ground allocation as much as climatic limitations, as demonstrated by deriving carbon budgets for plantations of *Pinus radiata* provided with different levels of available nitrogen.

For ecologists, the seasonal dynamics of carbon uptake and allocation are critical for they define the growing season, the time of flowering and seed production and, often, the susceptibility of forests to damage from frost, herbivores and pathogens. To meet this challenge, ecophysiolo-gists have applied insights developed from growth-room studies where photoperiod, nutrition, temperature and available water have been controlled. A number of empirical models have been developed to predict budbreak, foliage and stem elongation, and bud set in temperate and boreal forests from accumulated daily temperature values (heat sums) above a threshold after photoperiodic and chilling requirements have been met. In tropical and subtropical forests, the onset of a wet season (or the end of a dry season) may initiate conditions that can be predicted to favour the initiation or cessation of growth.

Although the theoretical basis for developing allocation models is still weak, some progress has been made. Various schemes have been proposed to model carbon allocation, as summarized by Cannell and Dewar (1994). These schemes are based on different assumptions about the controls on resource allocation as defined by: (i) sink strengths; (ii) resource deficiency ; (iii) functional balance ; (iv) distance from essential resources ; and (v) growth optimization. Although the various schemes on allocation are interrelated, none deals with underlying processes, and some are in direct conflict with one another.

Models that relate root uptake of nitrogen and shoot uptake of carbon have been improved by incorporating more details. Initially, growth was assumed to be a simple function of the amount of carbon and nitrogen available, with the product of carbohydrate and nitrogen resources in foliage and roots determining the relative allocation. Then models began to include resistance to transport and demands for resources by intermediate structures between roots and shoots. More recently, models consider sugar and amino acid transport separately through phloem and sapwood.

Root-to-shoot gradients in water potential also influence rates at which resources are transported through the two vascular systems. The majority of these models assume that growth at any instant is determined by the availability of substrate (C, N, etc.); however, availability is difficult to define. When plants are dormant, hormonal controls limit growth regardless of the concentration of carbohyrates and other resources in tissues. When shoot extension is rapid, root growth may be inhibited, even when starch concentrations are high.

Hormonal control is again involved in root development and there is also evidence that current photosynthate rather than starch reserves is required to produce new roots or to support the process of symbiotic nitrogen fixation. With the present state of understanding, it is reasonable to focus on developing accurate phenological models to define seasonal timing in growth activity as a function of daylength, temperature, drought and chilling requirements. Such phenological models would predict the fluctuating strength of competing sinks, while substrate availability would determine how much growth is possible.

With knowledge of phenology and environmental conditions, the pools of available carbon and nitrogen can be allocated into biomass, storage reserves and losses through respiration. Wein-stein *et al.* (1991) generated seasonal estimates of carbon allocation for red spruce *(Picea rubens)* by assuming priorities based on proximity to the resources (scheme iv in the list above). In this model, leaf growth has first priority for carbon after maintenance requirements of all living tissue are met. Storage of carbon in leaves is followed by growth and storage in branches, stem, coarse roots and, last, fine roots. Carbon in excess of that needed to meet the maximum growth rate of an organ was passed down the priority chain.

The priority ranking for allocation of water and nutrients would, according to this proximity logic, be opposite, with fine roots ranked first and new foliage ranked last. Similar proximity logic was discussed by Thornley (1976). Because critical resources come from opposite directions, no given organ is likely to grow at its full potential unless phenology limits all growth elsewhere. Predictions of photosynthate allocation seasonally in *Picea rubens* based on phenology and proximity logic showed reasonable growth patterns and matched measured storage reserves within 10% and total carbon content within 2% at the end of the year.

Most seasonal models of carbon allocation require specific knowledge of tree phenology, definition of the limits to growth of various organs, and specification of the size of storage reserves in all major organs. In reality, storage reserves should encompass not only starch but also amino acids, as well as a host of other compounds that protect against injury from temperature extremes and attacks from insects and diseases. Seasonally, trees are most susceptible to attack from insects and pathogens when storage reserves are low, a condition that occurs during rapid leaf expansion or when growth is prematurely halted during extended drought.

FUTURE CLIMATES AND THEIR SURFACES

To estimate the future climates for the decades centered about years 2030, 2060 and 2090, we updated the monthly normals of precipitation, minimum, maximum and average temperatures of all weather stations with outputs from the following General Circulation Models (GCMs) and scenarios: (a) Canadian Centre for Climate Modelling and Analysis (CCC), using the CGCM3 (T63 resolution) model, SRES A2 and B1 scenarios; (2) Met Office, Hadley Centre (HAD), using the HadCM3 model, SRES A2 and B2 scenarios; and (3) Geophysical Fluid Dynamics Laboratory (GFD), using the CM2.1 model, SRES A2 and B1 scenarios.

Data, their descriptions, and explanation of the scenarios are available at the International Panel on Climate Change Data Distribution Centre. Weather station records were updated by using a weighted average of the monthly change in climate calculated for the GCM cell centers lying within 400 km of a station.

The inverse of the square of the distance from the station to the cell centre was used for weighting. Of these emission scenarios, the A2 assumes high continued emissions from a continuously increasing population growth, with economic growth and technological change very heterogeneous among different regions and countries of the world; scenario B1 assumes a gradual reduction in emissions as rapid changes in economic structures are made toward a reduction in material intensity and introduction of clean technologies; scenario B2 assumes a continuous increasing population but at rate lower than scenario A2, intermediate levels of economic development and less rapid and more diverse technology change than B1 and A1. Scenario A1B assumes emissions intermediate between the A and B with a balanced fossil-intensive and non-fossil energy source in a world of very rapid economic growth as well as a rapid introduction of more efficient technologies.

The splines were then refit for each time period to produce monthly surfaces for the four climate variables for each scenario of all GCMs. Derived variables were then calculated as described above. Rather than updating grid cells of a fine resolution from the relatively coarse grids of the GCMs, our approach to downscaling begins anew the construction of spline surfaces from updated weather records. Either approach, however, tacitly assumes a constant relationship between elevation and the change in climate. Although this assumption is likely to be false, there are at present no reasonable alternatives. Our projections, therefore, are based on the differences in climate between that of a weather station and that of an average elevation of a GCM grid cell.

Bias would occur if these differences were dependent on the elevation of the weather station.

TOWARD UNDERSTANDING PLANT-CLIMATE RELATIONSHIPS

To illustrate the utility of the climate surfaces in biology, we consider (1) potential impacts of global warming on migration pattern of Mexican vegetation, (2) climatic niche analyses of a narrow endemic, *Pinus chiapensis*, (3) assisted migration in the botanical unique Tehuacán Valley as a management strategy for accommodating a changing climate, and (4) projection into future climate space the genetic differences occurring among populations within four species of pine inhabiting altitudinal transects in the Neovolcanic (also known as Transvolcanic) Axis. For the latter analyses, genetic responses of populations separated by 100 m of elevation have been studied previously for *Pinus oocarpa* populations from 1100 to 1500 m ; *P. devoniana* (also known as *P. michoacana*) populations from 1600 to 2400 m; *P. pseudostrobus* populations from 2100 to 2800 m ; and *P. hartwegii* populations from 3000 to 3600 m. Contemporary and 2030 values of five derived variables, MAT, MAP, DD5, MTCM and AAI, were estimated using the A2 scenario of CCC. From these estimates we calculate the altitudinal distance that populations would need to be transferred if they were to occur in a climate similar to that inhabited today. The underlying assumption is that existing populations are genetically adapted to contemporary climates.

To illustrate the use of the climate surfaces in climate niche analyses, we used the Random Forest classification tree of Breiman (2001) and followed the procedures detailed by Rehfeldt *et al.* (2006) to develop a statistical model to predict presence-absence of *P. chiapensis* from contemporary climate variables and to project the climate niche according to the 2060 climate of the A2 scenario of HAD. Breiman's algorithm develops a classification tree from two-thirds of the observations selected randomly from a data set and uses the remaining observations to calculate error. The programme then constructs a forest from a set of trees that use a recursive sample from the data set. For presence, we used all known locations inhabited by *P. chiapensis*, taken from Dvorak *et al.* (1996a), Newton *et al.* (2002), del Castillo and Trujillo (2008) and del Castillo *et al.* (2009), a total of 53. Because these observations constituted a census, we could assume that all other sites sampled from a digitized file of the Biotic Communities of North America would not be inhabited by the pine. Technical procedures, described in detail in Rehfeldt *et al.* (2006), include devising a sampling procedure according to which the number of observations taken from a community was determined by the size and number of polygons representing a community in the digitized file, procuring a systematic sample of observations from each polygon on the file, associating with each observation an elevation from the digitized elevation model of GLOBE (1999), and estimating the climate of each location from the spline surfaces. These procedures produced a pool of about 56,000 observations for which the pine was absent.

Because the Random Forests algorithm is best suited to data in which the number of observations in classes is approximately equal, only a small proportion of the number of observations without the pine could be used to construct a forest. In using the sampling protocol of Rehfeldt *et al.* (2006), we constructed 25 data sets, each with the 53 observations where the pine was present, weighted twice, and about 160 climatically diverse observations lacking the pine. Each data set was used in separate analyses to build a forest of 500 trees. This sampling protocol assures that 80 % of the observations without the pine will be among those for which separating presence from absence is the most difficult.

The programme started with 34 climate variables (19 derived variables previously described here plus additional interactions between them, such as DD5 x MAP) on which an iterative stepwise process eliminated one variable at each step according to the mean decrease in accuracy, a measure of variable importance. The process was halted with an eight-variable, shown previously to be robust for making projections. To make a prediction from the model, observation is run through all trees in all forests, with each tree contributing a vote, which in our case, would be whether or not the climate of an observation is suited for the pine. In making predictions for each cell of the GLOBE (1999) grid and, therefore, for gridded GCM projections, 12,500 votes were cast in each pixel. We assumed that the climate of a pixel was suited to the pine when a majority of the votes was affirmative.

PROJECTED CLIMATES AND THEIR SPLINE SURFACES

Weather station records updated for GCM output show in general that mean annual temperatures should increase steadily in Mexico, by 1.5 ºC in the decade surrounding 2030, 2.3 ºC in 2060, and 3.7 ºC by 2090. Projections, however, increasingly diverge among models and scenarios during the course of the century. For 2030, all the models and scenarios were similar, with the largest difference (0.5 ºC) between model CCC scenario A2 (increase of 1.7 ºC) and model GFD scenario B1 (increase of 1.2 ºC). By 2060, differences increased, with the largest difference (1.2 ºC) being between the most pessimistic scenario, A2 of HAD, which predicted a temperature increase of 2.8 ºC; the most optimistic was the B1 scenario of GFD which predicted an increase of 1.6 ºC. By year 2090, however, the differences among projections became even more pronounced, with model HAD scenario A2 projecting an increment of 5.0 ºC while GFD scenario B1 projecting 2.3 ºC. The 2090 projected increases shown in Figure are well within the range summarized for 21 global models for México and Central America for the period 2080 – 2099 Christensen *et al.* (2007).

The GCMs and their scenarios unanimously project a decrease in precipitation across the century, averaging -6.7% by 2030, -9% by 2060, and -18.2% for 2090. Variation among the GCMs and scenarios, however, was large. Projections for 2030, for instance, varied between +0.7 % and -13.5%, but for 2060, three of the projections predict increased precipitation in comparison to 2030, but with still an overall decrease from the present (ca. -3%), while four of the projections suggest a continued decline from 2030 to ca. -12% of the present. By 2090, all projections predict that precipitation should decrease, by -8.9% to -28.5% of the present. These results are similar to those of Christensen et al (2007) who calculated reductions ranging from -9 % to -48 % for México and Central America between 2080 and 2099 from 21 global models.

Differences projected for precipitation between the A and B scenarios increase considerably in time. Although the two scenarios purport a similar reduction in precipitation for 2030 of about -6.5 %, by 2060 the reduction for the A scenario is expected to be about -10.9 % while that for the B scenario is -5.7 %. By 2090, the reduction in precipitation under the A scenario (-22 %) is projected to be nearly twice that of the B scenario (-12.2 %). In using updated weather records to develop climate surfaces for the future, we examined 21 sets (seven model-scenarios by three time periods) of spline output statistics such as those of Table. Although the monthly means changed, the signal, RTMSE, and RTGCV remained similar. The fit of the models, therefore, was also similar.

MAPPED CLIMATE SURFACES

To illustrate the tremendous climatic variability in México, we mapped predicted mean annual temperature, mean annual precipitation, and the annual aridity index of the contemporary period and for 2090, the latter using the A2 scenario of CCC. As expected, geographic patterns in mean annual temperature reflect altitudinal differences between the mountain systems and the lowlands. In the contemporary climate, the coolest regions (<12 °C) are concentrated along the mountainous Sierra Madre Occidental and the central Neovolcanic Axis, where the highest Mexican volcanoes occur. Areas with > 25 °C occur along much of the coastal regions south of 25 °N as well as in the Yucatán Peninsula. Because the mean annual, maximum, and minimum temperatures and degree days > 5 °C are correlated, their geographic patterns are all similar. The 2090 temperature projections show that areas with the coolest climates (< 12 °C) should largely disappear, being restricted to only the highest volcanoes. In addition, portions of the Veracruz coast, Tabasco, Yucatán Peninsula, large parts of Pacific Coast and Balsas Depression (central-south of Michoacán State and western Guerrero State) became very warm, with average annual temperatures > 29 °C.

Highly variable contemporary precipitation results from both altitudinal effects and the differential impacts of arid westerly air masses from the Pacific

Ocean and moist monsoonal flows from the Gulf of México and the Caribbean Sea. In the Mediterranean climate of Baja California in northwest México, for example, annual precipitation may be only 200 mm, coming mostly in winter months. Yet, 3000 mm or more may fall at localities in the tropical rain forest along Tabasco and Veracruz slopes, northeastern Oaxaca and northern Chiapas in southeast México. This region is strongly influenced not only by monsoonal flows but also by occasional hurricane landfalls which contribute to the high variability in precipitation. Climate-change, however, is expected to progressively reduce the amount of area receiving more than 2300 mm of rain and expand arid and semiarid regions receiving < 400 mm (brown tones) in the north and northwest of México. Also, precipitation in much of the Yucatán Peninsula which currently receives 800 - 1400 mm should drop by about 17 %.

The annual aridity index (ratio of square root of degree days > 5 °C to precipitation) expresses an interaction of temperature with precipitation that better illustrates the remarkable climatic variability in México than either component separately. By reflecting the amount of growing season heat received for each mm of annual precipitation, this ratio represents the potential for moisture stresses to develop in the vegetation. In the contemporary climate, lowest index values are associated with the southern tropical forests, while the highest values occur in the deserts of the north. This map is strikingly similar to vegetation maps of México. However, according to the A2 scenario of CCC, the arid regions of north-central México, encompassing the States of Chihuahua, Durango, Coahuila, should expand toward both coasts and toward the southeast by 2090. At the same time, the moist regions of Veracruz, Tabasco and northern Oaxaca and Chiapas would be reduced greatly (blue tones), and much of the Sierra Madre Occidental and the Neovolcanic Axis would become more arid while the deserts of the northwest in Sonora and Sinaloa expand.

5

Assessment of Global Climate Change Management

CLIMATE CHANGE

Climate change outcomes and approaches to evaluation with different modelling systems were reviewed. Also, the interrelationships between the outcomes induced by climate change and other performance and structural characteristics of the models were suggested.

For example, in agricultural trade the underlying accompanying outcomes involve: domestic consumption and production systems and their regulation, macro economic and financial conditions and regulations and policies that indirectly effect these systems.

In this book, specific examples of models currently available for investigating impacts of climate change for agriculture and trade in agricultural commodities are considered. These include multi-market models and multi-sector or multinational models. To give the discussion added substance, specific modelling systems are identified, and their applicability is reviewed.

This review includes alternative methods for introducing climate impacts as well as the potential of the models for describing the consequences of the impacts of climate change, once introduced, on long and short-term trade patterns.

CARD/FAPRI MULTI-MARKET COMMODITY MODELLING SYSTEM

The modelling system is designated to include major agricultural commodities used for food, industrial, and animal production. It is dynamic and econometrically estimated from historical data. The countries explicitly modelled are primary exporters and importers of the indicated agricultural commodities. The system is currently being extended to include cotton and livestock and dairy products.

This modelling system was developed to investigate relatively short term impacts of government interventions and exogenous shocks in the international commodity markets. Exogenous shocks could include production and consumption levels reflecting simulated climate outcomes. The external shocks normally considered are from weather or from macro and financial policy.

Agricultural policies that have been studied include price stabilisation, trade restrictions, and domestic sector intervention. The opportunities for introducing policy change in this system are directly related to the importance of the included countries in international trade for the identified agricultural commodities. That is, the system is more policy specific for the more important trading countries, *e.g.*, the European Community for wheat and feed grains, Thailand for rice, etc.

ISSUES IN CLIMATE CHANGE

India should be an active and decisive partner, along with other developing countries, in climate change negotiations. We need to ask: What concentration levels, along with the associated risks, are acceptable to developing countries? How could it be ensured that the risks to the developing countries, and not just the costs to the developed countries, are minimised?

The assumptions about greenhouse gas (GHG) concentration levels for stabilisation of atmospheric concentrations range from 450, 550 and even 1000 ppmv. It should be noted that according to the IPCC third assessment report, increase of CO_2 concentrations in this range can lead to an equilibrium warming of between 2.0°C to 4.8°C.

The IPCC report probably specified a specific CO2 concentration increase with which the warming of 2 – 4.8 degrees was associated? Each of these concentration levels permits different reduction strategies. Freedom to choose options is more limited in the case of 450 ppmv as carbon budgets are very low and therefore greater mitigation action is required in the nearer term.

Integrated assessment models are currently being developed to consider these issues. Unfortunately, assumptions, premises and paradigms dictate the results of these models. In the view of the authors, often the developed countries' perspectives are hardwired into the models in such a manner that even if many scenarios are generated, the basic theme and results do not change.

For example, these models focus on minimising costs to the developed countries and not the risks to the developing countries. Yet, the decision about what is an acceptable level of climate change should centre around the risks to the developing countries. In fact, the developing countries should have a greater say as they are more vulnerable to the impact of climate change, they have a very small share in the cumulated emissions and thus have less responsibility for the problem of global warming; they are also poor and their emission trajectories are likely to rise due to development. In no other environmental issue are large polluters given opportunities to decide what cost and efforts are acceptable to them without full consideration of the vulnerability of the others. The level of effort needed to address an environmental issue is decided on the basis of what is good for society. For climate change, this will depend upon what risks of climate change impacts are associated with different levels of emissions.

India is a large country with wide ranging soil climate and other natural conditions. The results for India are thus likely to represent developing countries as a whole. Such risks to poor countries should be the primary focus of the climate change analysis, rather than costs to the developed countries.

To this extent, a paradigm shift is necessary from the cost-minimisation to risk- minimisation in the future analyses of IPCC. Of course, the nature and extent of climate change and its impacts are uncertain. That, however, should not be grounds for inaction. We should find a way to deal with differing perspectives on uncertainty and risk. We offer a suggestion for this later.

What are the implications of the differentiated responsibilities accepted under FCCC? If we look at the factors driving emissions, we get an idea. Parikh J modification of the Kaya identity states: Thus, carbon intensity of the energy system and energy intensity of GDP have to be also reduced to compensate for an increase in carbon due to an increase in GDP and population so as to reduce total emissions.

The Parikh identity also suggests that in the long-term population reduction, GDP stabilisation and other such measures that may be considered drastic by today's standards, will be in the arena of desirable options, if the climate change problem turns out to be more serious than it appears today and if through technology development we are unable to decouple carbon from energy use. Parikh J gives a detailed plan about step by step reduction for both developing and developed countries.

To stabilise or reduce carbon emissions in a smooth transition, one has to proceed in steps. Thus, first reduce the rate at which carbon emissions are growing, then make this rate zero, *i.e.*, stabilise carbon emissions and then make the rate negative, *i.e.*, reduce emissions.

For example, fossil fuel growth rates in OECD countries used to be in the range of 3% to 7% in the 70s, which came down in the range of 0% or ± 1%. On the other hand, emission growth rates in developing countries increased until the 1990s, but are showing signs of deceleration in many major countries such as India and China due both to reduced population growth as well as to reduction in energy intensity of GDP (E/Y and to some extent in carbon intensities (C/E) due to substitution of coal with oil and gas).

Yet these growth rates are at high level and further reduction in these growth rates is required after which they will also have to be stabilised of course, but that will take many decades.

This is a possible scenario. Carbon emissions in countries kept growing up to 1990 but at a decreasing rate. During the 1990's, emissions did not grow. From 2001 onwards, their emissions have to fall at an increasing rate. Carbon emissions will keep growing till 2050. The growth rate increases till 2000, remains stable from 2001 to 2025, and starts declining thereafter – becoming negative by 2050.

Discounting the Future

There is an implicit discount rate. It is often argued that the future generations would be richer and hence we can pass on the burden of emission mitigation on them. Some even suggested that they have to be similar to those used in any other investment strategy. It is often argued by some countries that we should use a high discount rate in designing climate change mitigation strategies. In our view, this is not quite correct.

Climate change, if it is permitted to happen, will impose a heavy burden on future generations in ALL countries not just on the citizens of countries. Even after 50 years, Indian nationals are likely to be poorer than those of the OECD are today.

Thus, by not taking actions now the burden is transferred not just to rich citizens of the OECD of the next generation but also to poorer Indians of tomorrow who would be poorer than today's citizens of OECD. A low discount rate is more appropriate when assessing optimal mitigation strategies.

Delay is Free Riding

Despite the commitments made at Rio, the countries have taken little action to meet their commitments. At the Conference of Parties (COP) in Kyoto in 1997, too, countries again delayed their commitments. On the whole the countries are expected to reduce their emission by 5.2% in the next fifteen years over their 1990 levels. Of this, the USA is expected to reduce by 7%, the EU by 8% and Japan by 6%. Even what little they agreed has still to be ratified in their home countries. Countries need to take urgent actions through a consensus-building exercise to engage local decision- makers. Through delays, rich OECD countries are occupying global environmental space.

During 1990 to 2020 (during which period they were supposed to act, haven't acted and are not likely to act) OECD countries would have emitted more than India would emit in the next 30 years, assuming a 5% increase in India's GHG emission every year. Point a shows the present emission level, and point c the target emission level in the year 2010.

The objective is to go from point a to point c. Path abc is the path that is likely to be followed if OECD countries were to fulfil their Kyoto obligations. Path adc, is the likely path of OECD emissions had they taken their FCCC commitment made at Rio seriously. While both the paths reach the same level, path abc puts much more CO2 in the atmosphere. The shaded area shows the additional CO_2 OECD countries have emitted and it would lead to higher temperature rise. One can recognise that this delay is costing India and other developing countries opportunities to develop in the future. Through delay OECD countries are further occupying global environmental space, and since Kyoto they have asked for even more of a delay. Delaying Kyoto is really reneging on Rio. To discourage free riding during the negotiation period and beyond, we suggest that countries are accountable for their own emissions for a specific period, say after 1990 or 2000.

That is, whatever decisions are arrived at, will be applicable retroactively from, say, 2000. That is, the clock starts ticking and all emissions are cumulated for each country even during negotiations. This way, negotiations will conclude faster and policy actions to reduce emissions will begin soon. Regardless of the outcome of the negotiations, these emissions will be shown against each country and that much less will be available to them in future. Thus, the countries taking actions in advance get their rewards and procrastinating countries will have to do more later.

Mitigation Costs and Benefits

Reluctance to take action now implies a faith in technical progress to effectively deal with climate change later. This is a risky strategy which the poor and vulnerable will find hard to accept. Technical progress sometimes brings with it unanticipated consequences. When CFCs where introduced, they were hailed as a great technical innovation – but, as it known now, they had the side-effect of leading to ozone depletion.

Thus, relying on technical solution alone can be risky. If dramatic technical progress does not take place, life style changes are inevitable if we want sustainable development. Also if you recognise the benefits, the costs of mitigation would not look too high. It shows a mitigation supply curve and a set of mitigation demand curves. If one is interested primarily in removal of distorting subsidies then D 1 may be the demand curve. If in addition, the society cares to control local air pollution, the relevant demand curve would be D 2, and so on to D 5, which adds an ethical dimension on species loss. A society's willingness would improve with greater awareness of its citizens. All countries need to put in efforts to increase it.

Benefits Justify Mitigation Cost

Demand for Abatement If primary concern is D 1 Removal of distorting subsidies D2D1 + local air pollution D3D2 + local externalities *e.g.* congestion D4D3 + ecosystem damage D5D4 + species loss.

The CDM – A Step towards Equity?

The Clean Development Mechanism (CDM) proposed in the Kyoto Protocol offers developing countries finance and technology by allowing countries to offset emissions through investing in emissions reduction in developing countries. Apart from the generally recognized problems of appropriate determination of the base line, India's concerns relate to getting fair compensation for sink projects, ensuring real transfer of technology and an uneasiness about selling 'low hanging fruits', *i.e.* the exploitation of cheap emissions reduction early on in the process by developed countries.

Sink Projects through CDM t is generally believed that the sink projects such as growing trees for afforestation and so on are some of the most attractive options. India has some 100 million hectares of wasteland and degraded forests on which such projects may be started. However, several major considerations may be important.

a) The trees fix carbon only during the growing periods. After reaching maturity, they are carbon- neutral. Thus, the carbon sink projects can create liabilities for the host country through committed land use.
b) If at the end of maturity forests are removed, it may appear in the statistics of land use change of India. If the wood is burnt, will the CO 2 generated be the liability of India?
c) If the forests are left intact, it may have implications if the opportunity costs of land become high in the meantime. However, only if the country had taken a careful long term decision to create green cover on a permanent basis, may such projects be considered. In the cases where the forest is removed or burnt, the global environment does not benefit as it would have merely postponed the problem. The liabilities are reduced only if energy crops are grown that will replace fossil fuels, for example, for wood based methanol, or wood-based power generation. However, who claims the credit, the one who supported the plantation or the one who uses it to replace fossil fuels?
d) Another difficulty relates to the measurement of carbon sequestered. This is not an easy task. All kinds of fudging are possible and there would be incentives to do so. One may also note that afforestation projects involve very little technology and hence very little technology transfer. Technology transfer is claimed to be a major advantage of the CDM.

Technology Transfer (TT) and CDM TT and CDM should be linked to ensure wider adoption of environmentally beneficial technologies beyond the CDM project. India would like to see that a "CDM project" leads to real technology transfer giving the country the ability not only to operate the technology but also to replicate and innovate.

Another concern of India is pricing of technology. There should be competition here. In a bilateral deal, the supplier of technology has monopoly power and the price charged for technology may be too high. Also projects such as sequestration projects do not involve technology transfer.

One way to ensure that CDM projects involve technology transfer at competitive prices is to require that every CDM project, including sequestration projects, make a specific contribution to a technology acquisition fund with which the developing country is free to buy technology not necessarily related to the CDM project, from anywhere in the world. This can moderate excessively high charges for technology from a monopolist supplier.

Low Hanging Fruits and Pricing of Carbon

When carbon is traded, what developing countries gain would depend on whether the market is competitive, whether futures markets exist, or whether the carbon is bilaterally traded in a project-by- project basis, as is envisaged under CDM.

Carbon Trading

As the developing countries have many low-cost opportunities to save GHG emissions (the low-hanging fruit) their marginal cost curve is relatively flat. Painuly has argued that developing countries are likely to get only about 20 % of the total surplus even under a competitive market. Should the developing countries then not opt for such trading? That would be an erroneous conclusion. If technical progress in the future lowers the demand drastically these low-hanging fruits would bring developing countries even less in future. The low-hanging fruits would then appear to just have rotten.

And in any case, since everybody tends to discount the future, money now is better than the same amount of money in future. However, developing countries can use these low-hanging fruits themselves in the future. So their long term opportunity cost may be higher than the short run marginal cost. To account for such opportunity costs, we should insist on the development of a futures market, so one can know how much the low-hanging fruits are going to be worth in the future.

Total Emission Reduction Commitment of A India is concerned that in bilateral negotiations between project parties, Indian entrepreneurs might only look at their private gain and sell carbon at throw-away prices, getting only AA'B. Developing countries should resist such trade. A well functioning market along with a futures market is the best way to ensure a good price. Development of such a market will take time. Meanwhile, a global carbon price floor should be announced for emissions trading and all developing countries should not trade below this price. India may want to do so unilaterally for its own projects.

Towards a Comprehensive Early Agreement

Disappointing delays in the implementation of the Kyoto Protocol lead us to suggest a system where all countries should be accountable for their cumulated emissions, say after 1990. When final negotiations are concluded, those countries that have taken early action will be rewarded and the others will have to do a lot more later. We suggest the following: Despite the uncertainties surrounding climate change, the risks of potentially adverse impacts on the food system, coastal zones and increased occurrence of extreme events should be avoided by early action. Even during the negotiation period, an immediate decision to work from cumulative emissions for each country from a given year, say 1990 or 2000, should be taken.

That is, whatever the final negotiated strategy, it will be applicable from the agreed reference year in the past to reward early actions by any country and perhaps to conclude negotiations faster. Delay to implement such a system only rewards current high emitters who do not take action to reduce emissions. Suppose we agree to limit climate change to 2 degree C of temperature increase. To give countries some leeway to deal with uncertainties involved and their differing perspectives of risk, countries must be held accountable for the damages caused due to their cumulative emissions over the most pessimistic scenario (*i.e.*, one which restricts the atmospheric CO2 level to the lower value for such a temperature increase).

However, a country may be permitted to emit up to their quotas as per the scenarios they consider reasonable. *Over time with research and better understanding of the global climate system the uncertainty will reduce and estimates of* the range of emissions *required* to restrict warming to 2C will narrow. Participating countries would be held responsible for the emissions that correspond to narrowed range. For greater flexibility quotas should be leaseable.

There are many desirable consequences of such a system. It will optimise response and reduce free-riding through delay. We have observed that the cost of delay in emission reduction (by the North) in terms of the South's foregone opportunities to development is substantial. This will impose many constraints on the way the South decides on policy options regarding issues such as how to generate power, how to use land, and what crops to grow and so on. Moreover, the South is highly vulnerable to the impact of climate change.

Hence, unless the North acts now, North-South transfers of large amounts will be needed to compensate the South for the development opportunities foregone or for direct economic losses stemming from climate change. The risks to poor countries should be the primary focus of the climate change analysis, rather than costs to the developed countries. To this extent, a paradigm shift is necessary from the cost- minimization in the future analyses of IPCC.

Our main arguments are as follows: India and other developing countries feel strongly that they are not responsible for the threat of climate change that has been created. Unsustainable consumption patterns of the rich industrialised nations in the world are responsible for it. Yet, India and other developing country economies may be highly vulnerable to climate change.

India's food production would be adversely affected. Sea level rise would displace a large number of people. The developing countries are particularly vulnerable to the likely increase in the incidence of extreme events. The impacts of climate change could hinder development and delay progress in eradicating poverty, potentially aggravating social and environmental conditions in these countries. An analysis of India's emissions show that its per capita emission of carbon is one fourth of the global average.

Even the top 10% of urban population emits well below the global average per capita emission. India, and other similar types of developing countries, are making significant progress in limiting GHG emissions through normal policy developments such as those aiming to improve energy and economic efficiency of the energy and industrial production capacity, as well as energy development, both conventional and renewable, which target improved environmental quality and limit human health hazards from air pollution.

India's energy intensity in industry and transport sector has come down. It has installed 2300 MW of generating capacity based on various renewables. Deforestation is arrested and the vast potential of afforestation on wasteland is increasingly utilised. India and many developing countries have carried out price reforms and removed subsidies.

These have resulted in substantial energy savings and reduction in emissions through greater use efficiency and fuel substitution. An equitable climate regime will focus on limiting the risks from climate change impacts to poor developing countries rather than on limiting the costs of mitigation per se. Options that improve economic efficiency of mitigation also need to address the distribution of economic costs associated with climate change.

Such a system needs to be guided by a better understanding of the potential economic impacts and other risk to developing countries which emanate from the climate change problem. One must also recognise the need for economic growth of developing countries. With differentiated responsibility, countries have to take the lead. For a smooth transition, they should first stabilise their carbon emissions as soon as possible and then reduce them to sustainable levels over the coming decades.

Emission of developing countries will need to grow even at increasing rates for some time. They would have to stabilize them somewhat later in the future and then reduce them. Unfortunately countries are delaying action. By their delay, they are occupying global environmental space and are free riding on developing countries. Compared to the carbon emission that OECD countries would have made had they followed the FCCC agreed on at Rio, they have emitted much more.

In fact, even if they were to meet the Kyoto targets by 2012, the additional emission of OECD countries between 1992 and 2012, exceeds the emission India is likely to make over 40 years assuming a 5% growth rate of emission. Delay cannot be justified on the ground that the future generations would be richer and should, therefore, bear higher costs of mitigation. Countries should use a low discount rate when assessing optimal mitigation strategies as postponing action now would put a larger burden of future population of developing countries – who would be poorer than what the citizens of countries are today. If countries recognise the environmental, societal and ecosystem benefits of mitigation and value them properly, it would justify incurring large mitigation costs. We need to increase awareness of citizens. One promising option for organising mitigation over the long term is to hold countries accountable for all emissions from some fixed date in time, say 1990 or 2000.

This provides incentives for early conclusion of negotiation. The CDM could be risky for developing countries because of perverse incentives to exaggerate valid credits (*e.g.* through exaggerated baselines) and because of likely imbalances in the power among the investor and host parties who will need to negotiate about important variables. To equitably share the gains from CDM projects, we may start with fixing a global carbon price floor. A major attraction of the CDM for developing counties is technology transfer.

However, carbon sequestration projects do not involve any significant technology. Also the price at which monopolist suppliers provide technology may vary. We suggest a technology acquisition fund in which every CDM project, including sequestration project, is required to make a contribution to technology funds with which technology catering to specific needs of developing countries can be developed (for example, 2 wheelers transport with 4 stroke engines or certain cheap cooling equipment).

Moreover, the developing country should be free to choose technology, not necessarily from the country that brings the CDM project, but from anywhere in the world. A more interesting option over the longer term could be to go to a fully fledged emission trading system, which would increase the economic efficiency of long term mitigation and, if emission quotas are allocated in an equitable way, begin to compensate developing countries for any costs that significant mitigation might impose on their developing economies.

Many persons in India and other developing countries are concerned about selling off their cheap mitigation options (the 'low-hanging fruits'). One should weigh the price one gets today, the worth of such fruits in the future and the possibilities of their 'rotting' if unused. The need for an approach to mitigating the threat of climate change that is equitable and one that can accommodate differing perspectives on risk need to be elaborated.

To initiate action now even with differing perspectives of uncertainties and risks that different countries have, we suggest a scheme where a global trading system of carbon emissions with futures market is introduced. The allocations of quotas are made on an equitable basis.

However, the total quota will depend upon each country's subjective trajectory that restricts global temperature change to a desired limit, say 2 degree C. Countries, however, are responsible for their cumulative emissions in carbon-ton-years that they have made and the range of permissible trajectories narrows as our knowledge and understanding improve. In the context of the current debate about climate change, it is necessary to show that far from being inactive, the developing countries, especially India, are taking considerable actions in terms of policies, programmes and projects. Technology transfer can speed up the modernisation process and additional funds can accelerate Government initiatives in energy conservation. However, policies for poverty alleviation must take priority. It is shown that savings in GHG emissions by the poor should not be expected at the expense of development.

CLIMATE CHANGE AND EAST ASIA

Bearing in mind the importance of East Asia for efforts to address global warming and climate change, what general lessons can we draw from this book's analyses of countries within the region? There are many, some rather broad and applicable to most countries and their relations with the world beyond their borders; others quite specific, often idiosyncratic, but which influence international relations in the region and beyond.

From a broad perspective (which we might think of as systemic forces), we must be cognizant of the environmental processes and effects themselves: global warming, resulting climatic changes, and the effects of these changes on the countries and peoples of East Asia. That is, climate change itself affects all other forces influencing the domestic and international politics of climate change in East Asia. Furthermore, *perceptions* of how climatic changes will affect national interests like economic development and human well-being are drivers of policy.

This highlights the important role played by science as the stimulus for political negotiations and policy responses, as well as the more general importance of knowledge and expertise (such as understandings of the economic impacts of climate change and possible responses to it) in shaping domestic and international policy.

Thus, at least at an initial level of analysis, countries of East Asia most affected by climate change (*e.g.* the Philippines and Indonesia) can be expected to take a great interest in international negotiations and mitigation measures.

However, there are other broad forces that influence political debates and policy processes. Among these are the disparities in wealth among countries, the different historical responsibilities for current atmospheric concentrations of GHGs, and associated considerations of international injustice and inequity.

These are very important concerns for the countries of East Asia, and they can divert attention away from the actual and perceived impacts of climate change, in the process politicizing the international policy process. Questions of international justice and equity highlight the important role for international assistance between richer and poorer countries.

For example, China has shown that it is possible to take concrete steps to address GHG emissions; its emissions, according to some new research, have not increased significantly in recent years, despite the rapid growth of its economy.

Even if its actual emissions are growing, they are not increasing at anywhere near the rate of economic growth. To be sure, this can be explained in large part by China's efforts at transition to less-polluting fuels for reasons that go beyond combating climate change. Nevertheless, China expects much more help from wealthier countries to help it use energy more efficiently and in cleaner forms, and it expects the developed countries to set an example before it takes even more action to specifically address atmospheric pollution related to climate change. Other developing countries in the region hold similar views.

Much of the financial aid to support the kinds of development in East Asia that are less harmful to the global atmosphere is coming from within the region; Japan has provided major development assistance to its neighbours. But the motivation for Japan's assistance is arguably not always, or even frequently, linked directly to global warming and climate change, let alone climate change justice.

Instead, Japan's policies are often associated more with the bureaucrats' perceptions of national and industrial interests, or aid for sustainable development is viewed as a way to bolster Japan's international standing. In addition to highlighting some of the nuances associated with North-South aid in the context of climate change, this shows that motivations for policies related to climate change are not as predictable and straightforward as cursory thinking would anticipate.

Broad considerations regarding climate change - many largely unrelated - therefore influence national policies. Looking more closely at individual cases, further lessons can be learned. For example, what role does the desire among countries for international reputation and leadership play? Some countries, China and Japan for example, want to be and are international leaders in various issue areas. But their desire to lead has restrictions.

China, for example, wants to lead the developing world, and in so doing it resists the demands of the industrialized countries to take on firm commitments to limit its GHG emissions, and it resists multilateral efforts to shape its emissions policies. It has, nevertheless, been successful in limiting its emissions voluntarily, and it readily joins with other countries (notably Japan) in bilateral efforts to move its economic development path in an environmentally sustainable direction.

This is because it can control such efforts more readily, take advantage of international financing associated with climate change, and otherwise promote its particularistic national interests more easily. Japan wants to lead as well, and it clearly cares about its image globally. But its climate change diplomacy may not be motivated greatly by the problem of climate change *per se*, which helps explain why its desire to lead is partly directed at shaping its regional neighbours' views of Japan and Japanese foreign policy.

Other lessons come from the case studies that follow. History and environmental experiences are sometimes crucial in international cooperation and policy making in East Asia. (Such findings are not new, to be sure, but they have been given too little treatment in existing literature addressing environmental diplomacy and sustainable development in the region.)

For example, the history of war and occupation, perhaps surprisingly, greatly impacts climate change policy in East and Southeast Asia. China remains weary of outside influences and pressures as a consequence of its domination by Western powers, particularly in the nineteenth century.

This affects its willingness - or rather unwillingness - to be bound by environmental standards set by international organizations, even if it has a hand in shaping those policies, and even more so limits its willingness to allow outsiders to dictate and run development projects related to climate change.

Furthermore, China and the countries of East Asia have not forgotten Japan's occupations and atrocities in the last century. Consequently, they expect Japan to provide aid as a form of retribution, and Japan has agreed to do so, often in the context of development assistance to combat pollution contributing to global warming (among other, more clearly self-interested, rationales for its aid).

Environmental history is also important. Japan has learned some important lessons from its national experiences with terrible environmental pollution, and this has affected its policies on climate change, both domestic and international. Thus it has had substantial success in addressing its environmental problems, although a fixation on domestic issues has often distracted it from taking on a more proactive role in dealing with global environmental problems.

China's experience with longstanding ecological scarcities and widespread environmental pollution is equally appalling, if not more so, especially considering the very large number of people adversely affected. This recent history, and its continuing manifestations, is pushing China toward much more action to combat the pollution that causes global warming, in so doing addressing some of the country's worst national ecological issues.

Other lessons that come from the research described in this book highlight the often highly pluralistic nature of climate change politics. In international forums, many actors are able to affect policy. The usual actors remain the most important, notably governments and their diplomatic representatives, as well as the international organizations and the officials working for them. But other actors are increasingly very important in international policy processes, notably international financial institutions, nongovernmental organizations, and commercial entities.

Similarly, a host of actors are at work at national and local levels. These range from state-level officials, bureaucracies, and local politicians, to a variety of actors in civil society, such as scientists, environmental activists and transnational groups, international organizations, and the communities and individuals most affected by climate change at the local level.

One interesting finding of some of the contributors to this volume is that the level of pluralism at the domestic level is not what it might at first appear to be. For example, Japanese democracy is shown to be often rather unresponsive to the interests of the Japanese people (that Japanese citizens have usually not tried harder to pressure government is of course an important consideration).

Thus, Japanese policy on climate change is largely the result of bargaining among state-level actors (*i.e.* bureaucrats and the ruling party) in association with Japanese industry. Perhaps surprisingly, however, China's climate change policy process is quite pluralistic. A myriad of actors at all levels of society are involved.

Communist Party officials are of course able to influence the shape of policy, but so too are scientists (many with conceptions of preferred policy derived from close contact with foreigners), bureaucrats of all stripes, local officials, and increasingly tolerated environmental activists - among many other actors.

These cases show that an understanding of climate change politics in East Asia, and the resulting policies, require us to look inside the "black box" of domestic politics, and also to look at how what goes on there is influenced by forces from the outside. What comes from them is a better understanding of East Asia's important role in ongoing international efforts to limit global warming and manage its effects, as well as the importance of considering how decisions taken farther afield affect the region and its peoples.

Hopefully they also raise new questions that will be explored by others, in so doing heightening awareness of how East Asia can and does contribute to this problem and, more importantly, to its solutions.

THE COSTS AND OPPORTUNITIES OF CLIMATE CHANGE IN SOUTHEAST ASIA

The concerns and policy-making challenges faced by the world's largest developing country: China. Joy V. Galvez helps us appreciate the impacts of climate change - and their implications for policy making - in a much smaller and arguably even more vulnerable developing country of East Asia: the Philippines.

Her summary of many key issues and concerns in the Philippines highlight some possible differences in perceived interests and strategies of the poorer countries in East Asia. As Galvez points out, the Philippines is a hot spot for natural hazards. The sectors in the Philippines projected to suffer the most from climate change-related impacts are water resources, agriculture, coastal resources, and human health. Galvez reminds us that these impacts worsen conditions in an already very underdeveloped country, and they will cause further suffering among a population that is already largely destitute.

Galvez shows how government and private actors have worked together to assess the Philippines' contributions to global warming and the implications of climate change for the country. At the international level, the Philippines was one of the first countries to discuss and develop positions on climate change. It aligned itself with other developing countries.

At the national level, the government developed an interagency group, composed of national agencies, academic institutions, and civil society actors, that has contributed to scientific research related to climate change vulnerability of smaller countries like the Philippines, as well as to the strong legal position of developing countries in relation to the responsibilities and commitments of developed countries. But Galvez believes that these actions have been inadequate.

Without a greater response from the national government, international organizations and the Filipino people, she says, the Philippines will suffer markedly from the projected impacts of climate change. Her chapter shows that if the Philippines is to cope with this problem, the government must develop programmes for massive information campaigns, educating the populace about the issue and how they contribute to it; review existing national and local environmental laws based on their relationship and congruence with international environmental laws, particularly on biodiversity conservation and climate change; and impose stiffer penalties on violators of existing environmental laws (*e.g.* strengthening the country's logging ban and clean air act) so as to curtail abuse and wanton environmental destruction.

Having said this, the necessary resources to actualize Galvez's recommendations are far too limited. Implementation of these strategies will require substantial additional assistance from the world's developed countries.

Frank Jotzo, Agus P. Sari, and Olivia Tanujaya use the case of Indonesia to demonstrate some possible fallacies in current thinking on international efforts to offset GHG emissions in developing countries. The Kyoto Protocol requires the developed industrialized countries to limit their GHG emissions as well as to enhance "sinks" for them (*e.g.* possibly including growing trees, which at least temporarily absorb carbon dioxide).

One of the protocol's unique characteristics is the provision for carbon emissions-offset mechanisms, where countries can trade emissions quota permits. At present, the CDM is the only mechanism for this that can include developing countries. The issues under negotiation include potential restrictions on buyers and sellers, and the inclusion of sink projects (largely forestry programmes) under the CDM.

Using a quantitative model developed specifically for policy issues in the implementation of the Kyoto Protocol (the Pelangi Emissions Trading model), Jotzo, Sari and Tanujaya analyze the implications of including sinks projects under the CDM. They find that assuming these projects will increase the volume of the CDM may be incorrect. This is because the increase in low-cost sinks projects leads to a fall in the price paid for emissions credits, which can outweigh the quantity gains, and lead to lower revenue and lower financial gains for developing countries. However, equity between countries and between regions within countries may be enhanced by the inclusion of sinks, as shown by the authors' study of Indonesia.

At the very least, their work shows that the "devil is in the details," and that proposals for dealing with climate change absent detailed analyses of their impacts may have unforeseen implications for social and international equity, as well as for the practical goal of reducing climate-changing emissions.

Tim Forsyth discusses the implications of climate change investment and technology transfer for countries in Southeast Asia. He argues that technology transfer is crucial to international environmental agreements, and that it is viewed by many developing countries as a prerequisite for their adherence to treaties.

Yet many investing countries see technology transfer as a lengthy and costly process that threatens intellectual property rights. Forsyth argues that such views need to be rethought. Technology transfer should instead be perceived in terms of so-called "horizontal" transfers (including long-term sharing of technological expertise) and "vertical" transfers (in which technologies are relocated without this long-term sharing).

Forsyth illustrates how vertical transfer may occur, using evidence from Thailand, Vietnam, Indonesia, and the Philippines. His key argument is that integrating technology transfer with international investment offers a powerful way to overcome disagreements in the climate change negotiations.

But, for this to happen technology transfer must be seen as a function of international investment and national and regional technology policy. If technology development is still seen in conventional terms as a linear process, to be controlled by indigenous companies, the prospect for enhancing international climate technology transfer is reduced because the process will be perceived as too costly and a risk to competitiveness.

However, if it is seen as a chance to invite new technology investment from international companies that do not expect to give up intellectual property rights, it is possible to have a win-win situation in which environmentally sound technology is increased, local development is assisted through the introduction of new investments, and investors are allowed into new markets.

Technology transfer can therefore fully complement both international environmental agreements and international private-sector investment. Redefining technology transfer away from the conventional view, which suggests that it can only assist potential economic competitors, has ensured that foreign policy objections have acted against moves to enhance technology transfer in the past? In contrast, viewing technology transfer in terms of "vertical transfer," or the relocation of economic activity without the sharing of intellectual property rights, can lead to an integration of foreign policy objectives with activities to mitigate climate change.

Seeing the relationship between technology transfer and other important aspects of foreign and economic policy may lead to more optimistic and successful negotiations under the FCCC. However, Forsyth cautions that there is a need for careful monitoring of all international investments under the CDM to ensure that new investments in technology actually reduce GHG emissions.

Understanding how and why governmental and nongovernmental actors cooperate at the international level and work at local and national levels to address global warming and climate change requires us to look at many actors and, ideally, to employ several levels of analysis.

The environmental foreign policies of the countries of East Asia, and specifically the interactions between their domestic politics and policy-making processes, on one hand, and international relations on global warming and climate change, on the other, are explained by the perceptions and actions of, and deliberations among, many often disparate actors at the individual, national, international, and global levels.

For example, powerful individuals in China can push policy in new directions; bureaucracies and industrial actors in Japan can shape regulations and policies on climate change; international cooperation can stimulate new actions at the national and local levels, or it can itself be shaped by actors at those levels; global forces, most notably the climate changes that now seem to be underway and the growing norm that countries ought to act to deal with them, are increasingly affecting the world's and East Asia's responses to global warming.

The upshot is that we need to look within the countries of East Asia, while simultaneously looking at interactions among them and between them and countries and other actors beyond the region, to fully explain their policies and to comprehend sufficiently how to bring about needed policy changes.

The chapters that follow should help, at least in a small way, to do this. If acted upon, the lessons from this book may help mitigate global warming in the long term and reduce the suffering that will result from the adverse impacts of climate change.

CLIMATIC FACTORS AND MODELLING CONSIDERATIONS

Review of the 19 studies illustrates that output from Global Climate Models models is the dominant source of data relating to climate change. In a few cases, the GCM output is supplemented by analogue data from past climate events. The Salinger and McKeon *et al.*, works use arbitrary shifts in climate as indicators of the extent of climate change. Another approach available is that of relying on solicited opinions of experts. Use of GCMs is popular because it insures consistency across geographic regions. The coarse spatial resolution of the GMC approach, however, is likely to require the refinement of GCM results by expert opinion, or the use of statistical methods to achieve climate parametre estimates at more economically-relevant geographic scales.

CLIMATIC FACTORS CONSIDERED

Temperature is the climate factor considered in all studies, closely followed by precipitation. Failure to include precipitation generally occurred when *1*) that factor was of limited importance to the agricultural sector considered, or *2*) because of the greater uncertainty of the relationship between climate change and precipitation. The potential for a CO_2 fertiliser effect was considered only in a few studies.

Timing of Climate Change

In all but one case, the timing of the climate change hypothesised was assumed to be instantaneous. Except for Pitovranov, *et al.*, analysis of the adjustment from the current climate regime to the specified alternative was not conducted. Extreme events, usually drought, were considered in a number of the analyses reviewed. However, year-toyear variability and the role of extreme events within those dynamics were not evaluated in these cases.

Geographic Unit of Analysis

Clearly the United States and Canada have been preferred regions for study. Although the size of the agricultural industry in those nations provides partial justification for focusing on them, much needs to be learned about AIACC in Asia, Africa, South America and Oceania. The study by Kane*et al.*, is unique within the group in that its focus is global in nature, emphasising the relationships between agricultural trade and climate change.

Modelling Unit Employed

The vast majority of studies utilise models at the level of the growing plant to estimate climate effects on yields and then incorporate that change in crop potential directly into a more aggregate model to estimate economic effects. The aggregation issue, long a source of error in such processes, is uniformly ignored.

Economic Model(s) Used

These include linear programming, quadratic programming, simulation, and econometric based models. The Kokoski and Smith effort incorporates a computable general equilibrium model and the study by Kane *et al.* uses an international trade model. Analysis at the firm level is either ignored, or is conducted using very simplistic techniques.

Motivation of the Decision-Maker

In nearly all cases, individual decision-maker motivations, which are at the root of potential adaptations to climate change, are not considered. At the aggregate level, standard assumptions about behaviour are incorporated except in those cases where the motivation of the decision maker is not specified. Often, this lack of specification reflects a "budgeting" approach to the aggregate analysis.

Assumptions about Changes in Non-climatic Factors

Evaluation of the interaction between climate change and key variables such as population growth and changes in resource availability are strikingly absent. In a few cases, the potential for technological change of some type is considered.

Decision-making Potentials

An analysis of potential adaptation by the individual producer is done in a number of cases. However, this adaptation process, which typically is limited to crop mix and a few management practices, is generally accomplished in a very rudimentary fashion.

Incorporation of Uncertainty Relative to the Extent of Climate Change

The existence of major uncertainties about the true nature of future climate change is a factor that is stressed in all the studies reviewed. Despite that concern, a large number of the studies do not include a specific analysis of that uncertainty.

In studies that do include this concern, the analysis is limited to fairly simplistic sensitivity modelling of only a few parametres.

Agricultural Activities Considered

The most common type of agricultural activity analysed is that of cereal crop production. In some cases, a range of commodities are considered, although the approach used to indicate crop yield and climate change relationships often is fairly general when multiple crops are considered.

A common practice is to base yield change relationships of a group of crops on analysis of climate effects of a single representative crop. Livestock is included in a number of efforts, but that inclusion rarely considers the direct effects of climate change on animal production. Instead, livestock effects are restricted to the demand side of the analysis.

Effects Measured

It is more informative to discuss the types of effects measured as a whole, rather than as a separate section for each of the separate categories. Nearly all studies explicitly measure the physical production effects of a change in climate. No study evaluates all six of the potential effects identified. Three studies produce measurements on four of the factors cited.

Comparative advantage, environmental effects and profitability are relatively popular factors to evaluate. It should be noted, however, that even though a single study may have considered a number of factors, that analysis may not have been accomplished in a completely integrated fashion. For example, the environmental effects may have been derived from detailed modelling at the field or watershed area. Profitability, however, may have been assessed using an aggregate model of a regional economy.

A striking result apparent is the paucity of studies that addressed impacts on food stocks. It was argued previously that food security and food sector economics are key driving forces fueling societal interest in this topic. If that is correct, the methods currently used for AIACC do not lead to results that pertain to the basic decision issues that society must address.

The text described specific characteristics of 19 empirical studies addressing the agricultural impacts of climate change. Although a necessary step toward an understanding of current methodological approaches, that analysis of specifics does not summarise very well the extent to which these studies have addressed the needs of policy-makers who are striving to understand climate change and its potential agricultural effects. It was suggested that a decision perspective should be used as the driving force for defining an appropriate methodology.

Types of decision-makers and the decision questions of likely relevance to them were identified. In summary form, three key decision issues dominate society's interest in AIACC:

- Food security (with both national and international dimensions);
- The economics of food consumption and production; and
- Investment alternatives in agricultural enterprises and institutions.

Providing relevant information about the future impacts of climate change on these three broad issues should drive the evaluation of the appropriateness of any methodological alternatives.

Deficiencies of Current Methodology

This schematic summarises the major deficiencies of current methodologies relative to the needs of decision-makers. Major sources of uncertainty are identified as relating to either climatic or non-climatic factors. With respect to AIACC, non-climatic factors can be further subdivided as:

1. Forces affecting the demand for agricultural production and
2. (Non-climatic) factors affecting the agricultural supply potential.

Relative to climatic factors, decision-makers are concerned with the impacts that policy actions to lessen the build-up of greenhouse gases will have. One is to indicate impacts if no policy responses are instituted, and the other is to indicate that policy measures are employed.

This entry relates to impact analyses that use current parametre values. Those rows and columns represent analyses providing information that would be useful to validate modelling procedures and to give perspective to decision-makers. Studies that employ only those parametres, however, do *not* directly address the issues faced by decision-makers.

The fundamental reason for this gap arises, not from the economic tools utilised, but rather from the underlying perspectives employed. Stated a bit differently, it is very unlikely that any analysis which is not focused on the correct question(s) will provide the most useful answer(s).

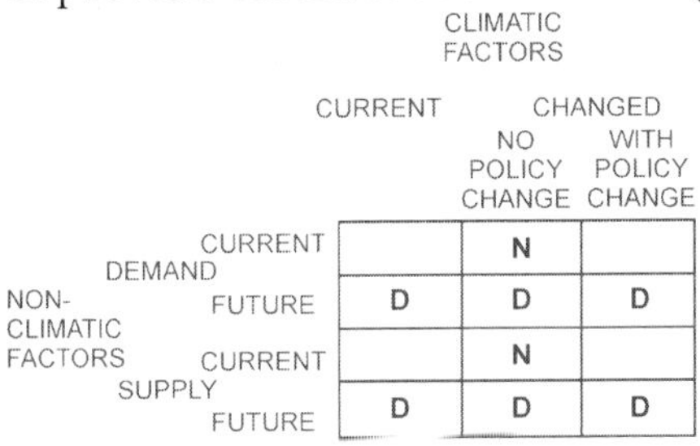

			CLIMATIC FACTORS	
		CURRENT	CHANGED	
			NO POLICY CHANGE	WITH POLICY CHANGE
NON-CLIMATIC FACTORS	DEMAND CURRENT		N	
	DEMAND FUTURE	D	D	D
	SUPPLY CURRENT		N	
	SUPPLY FUTURE	D	D	D

Fig.: Schematic of Information Needed toAddress Underlying Uncertainties

Future Directions

Methodology development must be recognised as being evolutionary in nature. Therefore, rather than suggesting that prior work was methodologically deficient, should be viewed as indicating needed future directions.

The remainder of this book will discuss three major thrusts which should underlie future analyses:

- Explicitly addressing changes in non-climatic factors;
- Investigating the dynamic aspects of a changing climate; and
- Broadening the scope of analysis.

Explicitly Addressing Changes in Non-climatic Factors

If food security concerns are a major justification for conducting future AIACCs, then future population levels and resource availability are just as important as changes in temperature or precipitation. In addition to population, key factors include the agricultural land base, economic growth, water availability, input costs, and socially-imposed restrictions on agricultural practices. Accurately predicting all these factors, of course, is an unreachable goal. Considering potential changes in these factors as they interrelate with the AIACC is a reasonable requirement, however. Further, it must be recognised that each of the reviewed studies has made an implicit assumption about all of these factors, and others. Each of these studies has implicitly assumed that factors would stay at their current levels forever. If there is one assumption that is clearly implausible, it is exactly the one that is used most often.

Significant changes in factors such as climate, population, and resource availability will not go unnoticed. Therefore, failure to explicitly consider decision-maker responses is just as serious as is ignoring the changes in the underlying factors themselves. Indeed, failure to explicitly consider these response functions significantly weakened the credibility of the AIACCs reviewed in this chapter.

The types of behavioural responses that seem to warrant consideration are:

- Adoption of alternative management practices that producers could employ using existing technologies;
- Changes in food consumption patterns, since consumers respond to differing supply regimes for food; and
- Efforts to create and implement technological improvements which mitigate or exploit alternative climatic regimes.

The process of considering non-climatic factors is interrelated with the incorporation of behavioural adaptations. For example, greater use of irrigation is one currently available management practice that might offset yield reductions due to adverse temperatures, or declines in natural rainfall levels. This response assumes that adequate irrigation water is economically available, a critical assumption. Failure to address uncertainties such as the availability of water for irrigation may send false signals about the resilience of the agricultural system.

Tracing the Path of a Changing Climate

Decision-makers live in a challenging, complex and constantly changing *present.* Analyses that abstract too far from that setting lessen their usefulness, no matter how technically well done. The year 2030 (or any other year 40 to 50 years into the future) may be an excellent reference point for technical and climate reasons.

From a decision perspective, particularly when the choice involves action to reduce the emission of greenhouse gases today, economic impact data tied to such a reference point loses meaning. Instead, the dynamics of changing conditions as society moves to that future point in time are of extreme relevance.

The technical feasibility of meaningfully tracing the impacts of a changing climate is a serious question. Our ability to understand and to realistically forecast both climate change and economic response functions over such long periods of time is quite limited.

However, it should also be kept in mind that such forecasts would not require the same precision that forecasting quarterly GNP for each of the next five years does. Instead, a broad understanding of the implication of climate change and societal adaptations are the ultimate goal. This (general) perspective may make the process achievable.

Addressing climate change as an evolving process opens tremendous opportunities. One is the explicit analysis of the potential for greater frequency of extreme climate events. Some analysts feel that greater variability may be a feature of a changing climate. If so, that variability may have profound, relatively immediate, implications for decision-makers. When food security and the economics of the global food system are key issues, variability of supply can be as economically relevant as are the average levels of these variables.

Broadening the Scope of Analysis

Analysis of prior studies displays a disappointing consistency of scope. Nearly all considered the impacts of climate change within the framework of cereal crops in developed nations. The economic importances of these activities are significant. But they do not comprise the majority of the world agricultural system. Therefore, there is a need to broaden the scope of analysis, in terms of geographic areas of analysis and agricultural activities included.

Africa, Asia and South America have virtually been ignored, yet from a global food security perspective they are critically important. Similarly the perennial crops, vegetables and livestock, of central importance in many nations, have not been considered. Data availability has been a key determinant driving the framing of analyses to date.

GLOBAL WARMING AND THE IMPACTS OF CLIMATE CHANGE

Particularly over the last decade, scientists have improved their understanding of the causes and consequence of global warming, which is the warming of the Earth from "greenhouse" gases (GHGs) trapped in the atmosphere. Many of these GHGs, such as carbon dioxide (CO2) and methane, while having natural sources, are also the products of human activities and industrialization. Carbon dioxide, which is the most influential GHG in aggregate, is created by the burning of fossil fuels (*e.g.* coal, oil, natural gas) for industry, transportation, and other purposes, and when trees are felled and subsequently decay or are burned. Other GHGs, such as methane, are the result of agriculture, and yet others (*e.g.* chlorofluorocarbons, which also deplete the stratospheric ozone layer) are released mostly by industrial activities. The Intergovernmental Panel on Climate Change (IPCC) has concluded that human activities are adding GHGs to the atmosphere, and that this additional contribution over natural sources is having a discernable impact by increasing global temperatures.

Climate change refers to the climatic changes and their consequences resulting from global warming, with the FCCC including under this rubric atmospheric changes connected directly or indirectly to human activities. The impacts of climate change on natural ecosystems and on human society and economies are potentially severe, ranging from sea-level rise and melting ice at higher latitudes (the Artic and Antarctic) and altitudes (mountain glaciers), to changing weather patterns characterized by increasingly severe storms, floods and droughts, and the attendant impacts of these changes, such as the spread of pests to warmed areas.

Some areas may experience positive effects of climate change (*e.g.* an extended growing season in high latitudes), but these will likely be accompanied by adverse impacts. Overall, predictions point to adverse impacts, particularly in parts of the world where geographic vulnerability and poverty make adaptation difficult or impossible. Much of East Asia falls into this latter area.

GLOBAL WARMING: CAUSES AND RESULTING CLIMATE CHANGE

The most authoritative reports on the causes and consequences of climate change come from the IPCC, particularly its 1995 Second Assessment Report (SAR) and its 2001 Third Assessment Report (TAR). The latter report refined the findings of the first assessment, pointing out that climate change is likely to be worse and occur more rapidly than initially predicted. Here I summarize the IPCC's findings on global warming and the worldwide effects of climate change before pointing out some of the anticipated socio-economic impacts in East Asia.

According to the IPCC's TAR, there is now a collective picture, derived from an increasing body of observations, of a warming world and other changes in the Earth's climate system. The global average surface temperature increased during the twentieth century, with the 1990s and early 2000s the warmest on record. Snow and ice cover have decreased, global average sea level has risen, and the heat content of the oceans has increased.

Other aspects of climate have changed during the twentieth century, including changes in precipitation (*e.g.* increased heavy precipitation events) and cloud cover; fewer extreme low-temperature periods and more high-temperature periods; more frequent, persistent, and intense episodes of the El Nino ocean-warming event (and related adverse effects on weather in many areas); and an increase in areas experiencing drought and severe wet periods.

Some climate related events, such as tornadoes or tropical storms, do not appear to have changed based on IPCC data, although the evidence is conflicting. The TAR also finds that emissions of GHGs from human activities are altering the atmosphere in ways that are expected to affect climate.

Human activities have increased atmospheric concentrations of GHGs (*e.g.* CO_2, methane, nitrous oxide, halocarbons) and their warming potential. According to the report, "atmospheric concentration of carbon dioxide (CO_2)

has increased 31 percent since 1750. The present CO_2 concentration has not been exceeded during the past 420,000 years and likely not during the past 20 million years. The current rate of increase is unprecedented during at least the past 20,000 years".

Three-quarters of human-induced emissions of CO_2 over the last two decades has come from the burning of fossil fuels (*e.g.* coal, oil, and natural gas), with most of the remainder the consequence of land-use changes, particularly deforestation. Natural causes of climate change have been relatively small. Furthermore, models for predicting future climate are increasingly accurate and precise. While uncertainties remain, understanding of climate processes and predicted effects has improved.

According to the TAR, new and stronger evidence points to human activities as the sources of observed global warming over the last fifty years, further strengthening the SAR's conclusion that the "balance of evidence suggests a discernible human influence on global climate". Warming over the last 100 years is unlikely to have been natural, with studies showing that global warming, particularly during the last 35-50 years, most likely resulted from human activities.

Thus, the TAR concludes: "In light of new evidence and taking into account the remaining uncertainties, most observed warming over the last fifty years is likely to have been due to the increase in GHG concentrations. Furthermore, it is very likely that the 20th century warming has contributed significantly to the observed sea level rise ... and widespread loss of land ice".

Furthermore, the TAR determined that human activities will continue to shape the Earth's atmosphere throughout this century and into the future, and average global temperatures and sea levels are projected to rise.

Emissions from burning fossil fuels will be the dominant source of atmospheric CO2 during this century. These emissions and those of other GHGs would have to be reduced to "a very small fraction of current emissions" to stabilize climate.

Global average temperature is projected by the IPCC to increase by 1.4-5.8 degrees Celsius during this century (more than anticipated in the SAR). This warming will occur at a rate faster than that observed in the twentieth century, "very likely to be without precedent during at least the last 10,000 years".

During this century, warming is expected to occur in most areas, but it should be particularly pronounced at northern high latitudes during winter.

Global mean sea level is expected to rise 0.09-0.88 metres in this century, with other very likely changes to include higher maximum temperatures and more hot days over most land areas, higher minimum temperatures and fewer cold days over most land areas, more intense precipitation events over many areas, increased summertime continental drying and drought over mid-latitude continental interiors, and more severe storms over some areas.

Ecological and Socio-economic Impacts of Climate Change

The ecological and socio-economic impacts of climate change are likely to be very significant and often painful. The TAR's findings on these impacts include the following: Regional changes in climate have already affected many physical and biological systems, with temperature increases being the most proximate cause.

Observed changes in regional climate have occurred in terrestrial, aquatic, and marine environments, and effects have included shrinking glaciers, thawing permafrost, reduced periods in which lakes and rivers are frozen, longer mid- and high-latitude growing seasons, shifts in animal and plant ranges to higher latitudes and altitudes, declines in populations of some animals and plants and reduced egg-laying in some birds, and insects populating new areas.

It appears that some social and economic systems have already been affected by increased floods and drought, but separating these ecological events from socio-economic factors is difficult.

The TAR shows that many human systems are sensitive to climate change, including water resources, agriculture, coastal zones and marine fisheries, settlements, energy, industry, financial services (*e.g.* insurance industries affected by increased claims), and human health.

Adverse impacts of climate change include reduced crop yields in most tropical and sub-tropical regions; decreased water availability in many water-scarce areas, especially the sub-tropics; more people exposed to increased mortality from heat stress and vector-borne diseases like cholera; widespread increase in flood risk from rising sea levels; and increasing demand for energy to cool areas affected by higher summer temperatures.

Some impacts may be positive, such as increased crop yields in some mid-latitude areas; potentially more timber if forests are managed appropriately (although increased pests could more than offset this); increased water availability for some water scarce areas; lower winter mortality in traditionally cold areas; and reduced winter demand for energy due to higher winter temperatures.

Many of the risks are unclear, and there is substantial potential for "large-scale and possibly irreversible impacts" from changing ocean currents, melting ice sheets, accelerated global warming due to atmospheric feedback effects, and so forth. In addition to efforts to mitigate climate change, the TAR argues that adaptation is a necessary strategy.

However, those people and societies with the least resources are most vulnerable because they are least able to adapt. Projected warming may result in a mixture of economic gains and losses for developed countries, but developing countries can expect mostly losses: "The projected distribution of economic impacts is such that it would increase the disparity in well-being between developed countries and developing countries," with the disparities increasing the greater the temperature increases. The upshot is that "More people are projected to be harmed than benefited by climate change," even if temperature increases are limited.

The TAR is not restricted to scientific and economic assessments. It argues that international justice and equity are important considerations when addressing climate change: "Inclusion of climatic risks in the design and implementation of national and international development initiatives can promote equity and development that is more sustainable and that reduces vulnerability to climate change".

Global Warming and Climate Change Impacts in East Asia

Clearly, the global effects of climate change are potentially major, and will likely lead to many adverse consequences, difficult choices, and expensive adaptation measures for much of the world's population. The countries of East Asia will not be immune to these changes, and in most cases will be among the worst affected due to their vulnerable geographies and economies.

Effects may not always be adverse, but even if they are not they will likely increase unpredictability and require adaptation. What are the expected impacts of climate change in East Asia? Several research reports have anticipated the effects of climate change for the region. Some of their findings are summarized here to convey the scale and nature of the potential changes.

According to a 1997 report from the IPCC on anticipated regional impacts of climate change, temperate Asia (including Japan, the Koreas, and most of China) has experienced an average annual temperature increase of more than 1 degree Celsius in the last century, mostly since the 1970s, with substantial warming expected in this century. Rainfall is expected to change in the area, with substantial declines expected in most of China (notably northern provinces). Permafrost in northeast China is expected to disappear (with release of methane, thus adding GHGs to the atmosphere) and glaciers will melt. Northern China is particularly vulnerable to expected changes in rainfall, exacerbating existing water shortages.

The area is likely to experience changing agricultural yields, with many crops likely to see reductions and a northward movement of crop zones and anticipated shortages of roundwood (partly due to increased demand). Delta coastlines in China "face severe problems" from sea-level rise, which will include salt water intrusion into aquifers. Japan will not be immune; already many parts of major coastal urban areas, with millions of residents, are below the mean high-water mark.

Providing protection for only some of these cities will cost tens of billions of dollars. Japan's beaches, which comprise about a quarter of its coastline, will be subject to erosion - and over half of existing beaches may disappear. Additionally, heat-related deaths throughout temperate Asia may increase sevenfold by the middle of this century.

The potential effects of climate change for tropical Asia (encompassing Southeast Asia) are also described in the IPCC's 1997 regional report. It points out that the region already suffers from increasing pollution, land degradation, and all manner of environmental problems resulting from rapid urbanization, industrialization, and economic development. Climate change will exacerbate these problems. In this area, mean temperatures have already gone up by 0.3-0.8 degrees Celsius over the last 100 years.

Forest cover will change as a consequence, possibly increasing, and forest types may change. Changes in evaporation and rainfall are likely to have detrimental effects on freshwater wetlands. Coastal areas will be most greatly affected by sea-level rise and increased ocean temperatures (the latter possibly preventing coral reefs from keeping up with sea-level rise). Mangrove and tidal wetlands will have difficulty adapting due to bordering infrastructure and human activities. Greater erosion, coastal flooding, and salinization of fresh water sources are probable. Delta regions of Southeast Asian countries are particularly vulnerable, and throughout this area several million people could be displaced by sea-level rise.

The costs of responding to the impacts of rising seas, in the words of the IPCC, could be immense. Glaciers feeding the area's rivers will melt, and there may be yearly reductions - albeit between seasonal flooding - in the flow of snow-fed rivers, adversely affecting agriculture, hydropower generation, and urban water supplies.

Agriculture will probably suffer (despite CO_2 fertilization) from temperature and moisture changes and possibly from increased pests, affecting, for example, wheat, rice, and sorghum crops (although much uncertainty, confounding planning, will obtain). According to the IPCC report, poor rural populations depending on traditional forms of agriculture or living on marginal lands are especially vulnerable. Increased vector-borne diseases such as dengue, malaria, and schistosomiasis will adversely affect human health in this area.

A 1999 report on climate change impacts prepared by Britain's Climatic Research Unit summarized many potential impacts for some of the countries of East Asia. In China, temperature increases are predicted to be greatest over northern areas, with changes in precipitation and threats to biodiversity. In Indonesia, forest fires are predicted to increase and endangered species may be threatened.

In Japan, heat waves will increase in frequency and intensity, coastlines and coastal infrastructure will be harmed, and reefs will be stressed. In the Philippines, rainfall will increase during the wet season and decrease during the dry season, reefs will suffer from warming water, and potentially millions of people will be threatened by sea-level rise.

Von Hippel has summarized a few of the possible impacts of climate change in Northeast Asia: pressure on agricultural resources and accelerated desertification leading to cross-border migrations, particularly from China to Russia; adverse climatic effects on North Korea's food production, possibly increasing military pressure on South Korea and creating economic burdens for reunification; increased demand for air conditioning, leading to higher fuel consumption and hence more local and regional air pollution (and adding still further to GHG emissions); salinization of breeding grounds for fish from sea-level rise, leading to reduced fishery yields that could exacerbate conflicts over marine resources; additional oil pollution (from shipments of oil imports) that may strain relations among countries sharing marine resources and shipping lanes; and increased economic costs from natural disasters like catastrophic storms, straining emergency and disaster relief resources in the region.

The 2001 TAR assessment of vulnerability in Asia shows that the region is potentially more susceptible to climate change than are some other regions of the world. It concludes that the developing countries of Asia are highly vulnerable to climate change, and their adaptability is low. (Developed countries of the region (*e.g.* Japan) are of course less vulnerable because they are more able to adapt to climate change.)

Floods, forest fires, cyclones, droughts and other extreme events have increased in temperate and tropical Asia. The TAR anticipates that while agricultural productivity could increase in northern parts of Asia, food security would suffer in arid, tropical, and temperate Asia due to reduced agricultural and aquaculture productivity from warmer water, sea-level rise, floods, droughts, and cyclones.

Water availability may decrease in arid and semi-arid Asia and possibly increase in northern Asia, and increased incidence of vector-borne diseases and heat-stress will threaten human health. Temperate and tropical Asia should anticipate increased rainfall and floods, and sea-level rise and more intense storms could "displace tens of millions of people in low-lying coastal areas of temperate and tropical Asia". Some parts of Asia will see climate change effects on transport, increased demand for energy, and adverse impacts on tourism. Land-use and land-cover changes will threaten biodiversity, and sea-level rise will adversely impact coral reefs and mangrove areas that are important for fisheries.

What comes from these (and other) reports on the impacts of climate change in East Asia is that many of the effects will be felt most by - and be most painful for - the developing countries of the region. They are generally more vulnerable and least able to cope due to poverty and existing environmental problems and resource scarcities.

A very large number of people throughout East Asia live in low-lying coastal regions, and they are threatened by sea-level rise, land subsidence, inundation of fresh water aquifers by salt water, and more frequent and violent storms from climate change. Island countries such as Indonesia and the

Philippines are especially vulnerable to climate change effects. They can expect freshwater shortages and damage to coastal areas and adjacent infrastructure, with concomitant adverse effects on tourism. (Indeed, in extreme cases it may one day be necessary for some small-island states to abandon their territory altogether. Representatives from these countries have for some time argued that they are *already* feeling the effects of rising oceans.

Other poorer countries in the region are vulnerable. For example, the World Bank reported that Chinese research has estimated that a 1-metre rise in sea level would inundate 92,000 square kilometres of China's coast, displacing 67 million people (and more as population increases). According to one assessment, future climate change will reduce soil moisture in China, particularly in the north, and this will increase the demand for agricultural irrigation, which will in turn add to existing severe water shortages.

In short, "Possible impacts of climate change on Chinese agriculture could be highly disruptive ...". Already vulnerable, China may also see greater weather extremes, including droughts in the north and floods in the south, and heat stroke and death will increase, as may occurrences of malaria, dengue fever, and other diseases. An ever-growing body of research shows that climate change will (and probably has already) adversely affected human health, and this is particularly true of East Asia. Southeast Asia is especially vulnerable to anticipated increasing incidence of vector-born diseases. Hotter weather will increase heat-related mortality in the region, as indicated by historical studies from China showing a strong correlation between peak summer temperatures and death rates.

But even the developed countries and regions of East Asia are unlikely to avoid harm from climate change. For example, while Japan's coastlines are not as vulnerable as those of China, the Philippines, and other countries, it is reasonable to expect that it will suffer costly damage from sea-level rise, associated storm surges, and adverse weather, and it has direct interests in the health of surrounding seas and indirect interest in what happens throughout the region. By way of example, recently reduced fish catches by Japanese fisherman have been attributed to changes in underwater currents triggered by global warming. And there will be adverse impacts for Japan's biodiversity, forests, agriculture, wetlands, and water systems, as well as for infrastructure and human health. According to the government, climate change effects have already become visible in Japan. For these and other reasons, Japan supports the climate change regime and the Kyoto Protocol- despite its greater ability to cope with climate change compared to its neighbours.

GLOBAL FOOD SECURITY UNDER CLIMATE CHANGE

The Food and Agriculture Organization (FAO) defines food security as a "situation that exists when all people, at all times, have physical, social, and economic access to sufficient, safe, and nutritious food that meets their dietary needs and food preferences for an active and healthy life" (1). This definition comprises four key dimensions of food supplies: availability, stability, access, and utilization.

The first dimension relates to the availability of sufficient food, *i.e.*, to the overall ability of the agricultural system to meet food demand. Its subdimensions include the agro-climatic fundamentals of crop and pasture production (2) and the entire range of socio-economic and cultural factors that determine where and how farmers perform in response to markets. The second dimension, stability, relates to individuals who are at high risk of temporarily or permanently losing their access to the resources needed to consume adequate food, either because these individuals cannot ensure *ex ante* against income shocks or they lack enough "reserves" to smooth consumption *ex post* or both.

An important cause of unstable access is climate variability, *e.g.*, landless agricultural laborers, who almost wholly depend on agricultural wages in a region of erratic rainfall and have few savings, would be at high risk of losing their access to food. However, there can be individuals with unstable access to food even in agricultural communities where there is no climate variability, *e.g.*, landless agricultural laborers who fall sick and cannot earn their daily wages would lack stable access to food if, for example, they cannot take out insurance against illness. The third dimension, access, covers access by individuals to adequate resources (entitlements) to acquire appropriate foods for a nutritious diet. Entitlements are defined as the set of all those commodity bundles over which a person can establish command given the legal, political, economic, and social arrangements of the community of which he or she is a member. Thus a key element is the purchasing power of consumers and the evolution of real incomes and food prices. However, these resources need not be exclusively monetary but may also include traditional rights, *e.g.*, to a share of common resources. Finally, utilization encompasses all food safety and quality aspects of nutrition; its subdimensions are therefore related to health, including the sanitary conditions across the entire food chain. It is not enough that someone is getting what appears to be an adequate quantity of food if that person is unable to make use of the food because he or she is always falling sick.

Agriculture is not only a source of the commodity food but, equally importantly, also a source of income. In a world where trade is possible at reasonably low cost, the crucial issue for food security is not whether food is "available," but whether the monetary and nonmonetary resources at the disposal of the population are sufficient to allow everyone access to adequate quantities of food. An important corollary to this is that national self-sufficiency is neither necessary nor sufficient to guarantee food security at the individual level. Note that Hong Kong and Singapore are not self-sufficient (agriculture is nonexistent) but their populations are food-secure, whereas India is self-sufficient but a large part of its population is not food-secure.

Numerous measures are used to quantify the overall status and the regional distribution of global hunger. None of these measures covers all dimensions and facets of food insecurity described above. This also holds for the FAO indicator of undernourishment, the measure that was used in essentially all studies reviewed in this article. The FAO measure, however, has a number of advantages. First, it covers two dimensions of food security, availability and access; second, the underlying methodology is straightforward and transparent; and third, the parameters and data needed for the FAO indicator are readily available for past estimates and can be derived without major difficulties for the future.

This article reviews recent studies that have quantified the impacts of climate change on global food security. It starts with an overview of the principal aspects of climate change and their impacts on the four dimensions of food security. It then reviews model-based results and discusses the main findings that have arisen from these assessments. Finally, limitations of the current modeling system of potential surprises and some suggestions to improve future assessments to enhance their overall robustness and their relevance for policy makers.

CLIMATE CHANGE AND FOOD SECURITY

Impacts on Food Production and Availability

Climate change affects agriculture and food production in complex ways. It affects food production directly through changes in agro-ecological conditions and indirectly by affecting growth and distribution of incomes, and thus demand for agricultural produce. Impacts have been quantified in numerous studies and under various sets of assumptions. A selection of these results is presented in *Quantifying the Impacts on Food Security*. Here it is useful to summarize the main alterations in the agro-ecological environment that are associated with climate change.

Changes in temperature and precipitation associated with continued emissions of greenhouse gases will bring changes in land suitability and crop yields. In particular, the Intergovernmental Panel on Climate Change (IPCC) considers four families of socio-economic development and associated emission scenarios, known as Special Report on Emissions Scenarios (SRES) A2, B2, A1, and B1. Of relevance to this review, of the SRES scenarios, A1, the "business-as-usual scenario," corresponds to the highest emissions, and B1 corresponds to the lowest. The other scenarios are intermediate between these two. Importantly for agriculture and world food supply, SRES A2 assumes the highest projected population growth of the four (United Nations high projection) and is thus associated to the highest food demand. Depending on the SRES emission scenario and climate models considered, global mean surface temperature is projected to rise in a range from 1.8°C (with a range from 1.1°C to 2.9°C for SRES B1) to 4.0°C (with a range from 2.4°C to 6.4°C for A1) by 2100. In temperate latitudes, higher temperatures are expected to bring predominantly benefits to agriculture: the areas potentially suitable for cropping will expand, the length of the growing period will increase, and crop yields may rise.

A moderate incremental warming in some humid and temperate grasslands may increase pasture productivity and reduce the need for housing and for compound feed. These gains have to be set against an increased frequency of extreme events, for instance, heat waves and droughts in the Mediterranean region or increased heavy precipitation events and flooding in temperate regions, including the possibility of increased coastal storms; they also have to be set against the fact that semiarid and arid pastures are likely to see reduced livestock productivity and increased livestock mortality.

In drier areas, climate models predict increased evapotranspiration and lower soil moisture levels. As a result, some cultivated areas may become unsuitable for cropping and some tropical grassland may become increasingly arid. Temperature rise will also expand the range of many agricultural pests and increase the ability of pest populations to survive the winter and attack spring crops.

Another important change for agriculture is the increase in atmospheric carbon dioxide (CO_2) concentrations. Depending on the SRES emission scenario, the atmospheric CO_2 concentration is projected to increase from H"379 ppm today to >550 ppm by 2100 in SRES B1 to >800 ppm in SRES A1FI. Higher CO_2 concentrations will have a positive effect on many crops, enhancing biomass accumulation and final yield. However, the magnitude of this effect is less clear, with important differences depending on management type (*e.g.*, irrigation and fertilization regimes) and crop type. Experimental yield response to elevated CO_2 show that under optimal growth conditions, crop yields increase at 550 ppm CO_2 in the range of 10% to 20% for C_3 crops (such as wheat, rice, and soybean), and only 0–10% for C_4 crops such as maize and sorghum. Yet the nutritional quality of agricultural produce may not increase in line with higher yields. Some cereal and forage crops, for example, show lower protein concentrations under elevated CO_2 conditions.

Finally, a number of recent studies have estimated the likely changes in land suitability, potential yields, and agricultural production on the current suite of crops and cultivars available today. Therefore, these estimates implicitly include adaptation using available management techniques and crops, but excluding new cultivars from breeding or biotechnology. These studies are in essence based on the FAO/International Institute for Applied Systems Analysis (IIASA) agro-ecological zone (AEZ) methodology. For instance, pioneering work in ref. 9 suggested that total land and total prime land would remain virtually unchanged at the current levels of 2,600 million and 2,000 million hectares (ha), respectively. The same study also showed, however, more pronounced regional shifts, with a considerable increase in suitable cropland at higher latitudes (developed countries +160 million ha) and a corresponding decline of potential cropland at lower latitudes (developing countries "110 million ha). An even more pronounced shift within the quality of cropland is predicted in developing countries. The net decline of 110 million ha is the result of a massive decline in agricultural prime land of H"135 million ha, which is offset by an increase in moderately suitable land of >20 million ha. This quality shift is also reflected in the shift in land suitable for multiple cropping. In sub-Saharan Africa alone, land for double cropping would decline by between 10 million and 20 million ha, and land suitable for triple copping would decline by 5 million to 10 million ha. At a regional level, similar approaches indicate that under climate change, the biggest losses in suitable cropland are likely to be in Africa, whereas the largest expansion of suitable cropland is in the Russian Federation and Central Asia.

Impacts on the Stability of Food Supplies

Global and regional weather conditions are also expected to become more variable than at present, with increases in the frequency and severity of extreme events such as cyclones, floods, hailstorms, and droughts. By bringing greater fluctuations in crop yields and local food supplies and higher risks of landslides and erosion damage, they can adversely affect the stability of food supplies and thus food security.

Neither climate change nor short-term climate variability and associated adaptation are new phenomena in agriculture, of course. As shown, for instance, in ref. 9, some important agricultural areas of the world like the Midwest of the United States, the northeast of Argentina, southern Africa, or southeast Australia have traditionally experienced higher climate variability than other regions such as central Africa or Europe. They also show that the extent of short-term fluctuations has changed over longer periods of time. In developed countries, for instance, short-term climate variability increased from 1931 to 1960 as compared with 1901 to 1930, but decreased strongly in the period from 1961 to 1990. What is new, however, is the fact that the areas subject to high climate variability are likely to expand, whereas the extent of short-term climate variability is likely to increase across all regions. Furthermore, the rates and levels of projected warming may exceed in some regions the historical experience.

If climate fluctuations become more pronounced and more widespread, droughts and floods, the dominant causes of short-term fluctuations in food production in semiarid and subhumid areas, will become more severe and more frequent. In semiarid areas, droughts can dramatically reduce crop yields and livestock numbers and productivity. Again, most of this land is in sub-Saharan Africa and parts of South Asia, meaning that the poorest regions with the highest level of chronic undernourishment will also be exposed to the highest degree of instability in food production. How strongly these impacts will be felt will crucially depend on whether such fluctuations can be countered by investments in irrigation, better storage facilities, or higher food imports. In addition, a policy environment that fosters freer trade and promotes investments in transportation, communications, and irrigation infrastructure can help address these challenges early on.

Impacts of Climate Change on Food Utilization

Climate change will also affect the ability of individuals to use food effectively by altering the conditions for food safety and changing the disease pressure from vector, water, and food-borne diseases. The IPPC Working Group II provides a detailed account of the health impacts of climate change of its fourth assessment report. It examines how the various forms of diseases, including vector-borne diseases such as malaria, are likely to spread or recede with climate change. This article focuses on a narrow selection of diseases that affect food safety directly, *i.e.*, food and water-borne diseases.

The main concern about climate change and food security is that changing climatic conditions can initiate a vicious circle where infectious disease causes or compounds hunger, which, in turn, makes the affected populations more susceptible to infectious disease. The result can be a substantial decline in labour productivity and an increase in poverty and even mortality. Essentially all manifestations of climate change, be they drought, higher temperatures, or heavy rainfalls have an impact on the disease pressure, and there is growing evidence that these changes affect food safety and food security.

The recent IPCC report also emphasizes that increases in daily temperatures will increase the frequency of food poisoning, particularly in temperate regions. Warmer seas may contribute to increased cases of human shellfish and reef-fish poisoning (ciguatera) in tropical regions and a poleward expansion of the disease. However, there is little new evidence that climate change significantly alters the prevalence of these diseases. Several studies have confirmed and quantified the effects of temperature on common forms of food poisoning, such as salmonellosis. These studies show an approximately linear increase in reported cases for each degree increase in weekly temperature. Moreover, there is evidence that temperature variability affects the incidence of diarrhoeal disease. A number of studies found that rising temperatures were strongly associated with increased episodes of diarrhoeal disease in adults and children. These findings have been corroborated by analyses based on monthly temperature observations. Several studies report a strong correlation between monthly temperature and diarrhoeal episodes on the Pacific Islands, Australia, and Israel.

Extreme rainfall events can increase the risk of outbreaks of water-borne diseases particularly where traditional water management systems are insufficient to handle the new extremes. Likewise, the impacts of flooding will be felt most strongly in environmentally degraded areas, and where basic public infrastructure, including sanitation and hygiene, is lacking. This will raise the number of people exposed to water-borne diseases (*e.g.*, cholera) and thus lower their capacity to effectively use food.

Impacts of Climate Change on Access to Food

Access to food refers to the ability of individuals, communities, and countries to purchase sufficient quantities and qualities of food. Over the last 30 years, falling real prices for food and rising real incomes have led to substantial improvements in access to food in many developing countries. Increased purchasing power has allowed a growing number of people to purchase not only more food but also more nutritious food with more protein, micronutrients, and vitamins. East Asia and to a lesser extent the Near-East/ North African region have particularly benefited from a combination of lower real food prices and robust income growth. From 1970 to 2001, the prevalence of hunger in these regions, as measured by FAO's indicator of undernourishment, has declined from 24% to 10.1% and 44% to 10.2% respectively. In East Asia, it was endogenous income growth that provided the basis for the boost in demand for food, which was largely produced in the region; in the Near-East North African region demand was spurred by exogenous revenues from oil and gas exports, and additional food supply came largely from imports. But in both regions, improvements in access to food have been crucial in reducing hunger and malnutrition.

FAO's longer-term outlook to 2050 suggests that the importance of improved demand side conditions will even become more important over the next 50 years. The regions that will see the strongest reductions in the prevalence of undernourishment are those that are expected to see the highest rates of income growth. South Asia stands to benefit the most. Spurred by high income growth the region is expected to reduce the prevalence of undernourishment from >22% currently to 12% by 2015 and just 4% by 2050. Progress is also expected for sub-Saharan Africa, but improvements will be less pronounced and are expected to set in later. Over the next 15 years, for instance, the prevalence of undernourishment will decline less than in other regions, from H"33% to a still worrisome 21%, as significant constraints (soil nutrients, water, infrastructure, etc.) will limit the ability to further increase food production locally, while continuing low levels of income rule out the option of importing food. In the long run, however, sub-Saharan Africa is expected to see a more substantial decline in hunger; by 2050, <6% of its total population are expected to suffer from chronic hunger. It is important to note that these FAO projections do not take into account the effects of climate change.

By coupling agro-ecologic and economic models, others have gauged the impact of climate change on agricultural gross domestic product (GDP) and prices. At the global level, the impacts of climate change are likely to be very small; under a range of SRES and associated climate-change scenarios, the estimates range from a decline of "1.5% to an increase of +2.6% by 2080. At the regional level, the importance of agriculture as a source of income can be much more important. In these regions, the economic output from agriculture itself (over and above subsistence food production) will be an important contributor to food security. The strongest impact of climate change on the economic output of agriculture is expected for sub-Saharan Africa, which means that the poorest and already most food-insecure region is also expected to suffer the largest contraction of agricultural incomes. For the region, the losses in agricultural GDP, compared with no climate change, range from 2% to 8% for the Hadley Centre Coupled Model, version 3 and 7–9% for the Commonwealth Scientific and Industrial Research Organisation projections.

Impacts on Food Prices

Essentially all SRES development paths describe a world of robust economic growth and rapidly shrinking importance of agriculture in the long run and thus a continuation of a trend that has been underway for decades in many developing regions. SRES scenarios describe a world where income growth will allow the largest part of the world's population to address possible local production shortfalls through imports and, at the same time, find ways to cope with safety and stability issues of food supplies. It is also a world where real incomes rise more rapidly than real food prices, which suggests that the share of income spent on food should decline and that even high food prices are unlikely to create a major dent in the food expenditures of the poor. However, not all parts of the world perform equally well in the various development paths and not all development paths are equally benign for growth. Where income levels are low and shares of food expenditures are high, higher prices for food may still create or exacerbate a possible food security problem.

There are a number of studies that have ventured to measure the likely impacts of climate change on food prices. The basic messages that emerge from these studies are: first, on average, prices for food are expected to rise moderately in line with moderate increases of temperature ; some studies even foresee a mild decline in real prices until 2050. Second, after 2050 and with further increases in temperatures, prices are expected to increase more substantially. In some studies and for some commodities (rice and sugar) prices are forecast to increase by as much as 80% above their reference levels without climate change. Third, price changes expected from the effects of global warming are, on average, much smaller than price changes from socio-economic development paths. For instance, the SRES A2 scenario would imply a price increase in real cereal prices by H"170%. The (additional) price increase caused by climate change (in the Hadley Centre Coupled Model, version 3, climate change case) would only be 14.4%. Overall, this appears to be the sharpest price increase reported and it is not surprising that this scenario would imply a stubbornly high number of undernourished people until 2080. However, it is also needless to say that a constant absolute number of undernourished people would still imply a sharp decline in the prevalence of hunger, and, given the high population assumptions in the SRES A2 world (>13.6 billion people globally and >11.6 billion in the developing world) this would imply a particularly sharp drop in the prevalence from currently 17% to H"7% by 2080.

QUANTIFYING THE IMPACTS ON FOOD SECURITY

A number of studies have recently quantified the impacts of climate change on food security. In terms of quantifying agronomic yield change projections, these studies are either based on the AEZ tools developed by the IIASA analysis or the Decision Support System for Agrotechnology Transfer suite of crop models; all use the IIASA-Basic Linked System (BLS) economic model for assessing economic impacts. These tools, with some modifications relating to how crop yield changes are simulated, have also been used by others to undertake similar assessments and provide sensitivity analyses across a range of SRES and general circulation model (GCM) projections. Many other simulations have also examined the effects of climate change with and without adaptation (induced technological progress, domestic policy change, international trade liberalization, etc.), with and without mitigation (*e.g.*, CO_2 stabilization, variants for temperature, rainfall change and distribution) or provide impact assessments for different speeds of climate change. The quantitative results for food security, trying to illuminate some of the differences and extract the main messages that emerge from the various studies. Unless indicated, all simulation results discussed below include the combined effects of climate change and elevated CO_2 on crops. The key messages can be summarized as follows:

First, it is very likely that climate change is likely to increase the number of people at risk of hunger compared with reference scenarios with no climate change; the exact impacts will, however, strongly depend on the projected socio-economic developments. For instance, it is estimated that climate change would increase the number of undernourished in 2080 by 5–26%, compared with no climate change or by between 5 million and 10 million (B1 SRES) and 120 million to 170 million people (A2 SRES), with within-SRES ranges depending on GCM climate projections. Using only one GCM scenario, others projected small reductions by 2080, *i.e.*, –5%, or –10 million (B1) to –30 million (A2) people, and slight increases of +13%–26%, or H–10 million (B2) to 30 million (A1) people.

Second, it is likely that the magnitude of these climate impacts will be small compared with the impact of socio-economic development. The limitations of socio-economic forecasts, these studies suggest that robust economic growth and a decline in population growth projected for the 21st century will, in all but one scenario (SRES A2), significantly reduce the number of people at risk of hunger in 2080. At any rate, the prevalence of undernourishment will decline as all scenarios assume that world population will continue to grow to 2080, albeit at lower rates. Compared with FAO estimates of 820 million undernourished in developing countries today, several studies estimate reductions of >75% by 2080, or H–560 million to 700 million people, projecting 100 million to 240 million undernourished by 2080 (A1, B1, and B2). The only exception is scenario A2, where the number of the hungry is forecast to decrease only slightly by 2080; but the higher population growth rates in A2 compared with other scenarios mean that the prevalence of undernourishment will decline drastically. However, these analyses also confirm that the progress will be unevenly distributed over the developing world, and more importantly progress will be slow during the first decades of the outlook. With or without climate change, the millennium development goal of halving the prevalence of hunger by 2015 is unlikely to be realized before 2020–2030.

In addition to socio-economic pressures considered within the IPCC SRES scenarios, food production may increasingly compete with bio-energy in coming decades; studies addressing possible consequences for world food supply have only started to surface, providing both positive and negative views. Importantly, none of the major world food models discussed herein have yet considered such competition.

Third, sub-Saharan Africa is likely to surpass Asia as the most food-insecure region. However, this is largely independent of climate change and is mostly the result of the socio-economic development paths assumed for the different developing regions in the SRES scenarios. Throughout most SRES and climate-change scenarios sub-Saharan Africa accounts for 40–50% of global hunger by 2080, compared with H–24% today; in some simulations sub-Saharan Africa even accounts for 70–75% of global undernourishment by 2080. These high estimates have emerged from slower growth variants of the A2 and B2 scenarios; also an A2 variant with slower population growth yields a sharper concentration of hunger in sub-Saharan Africa. For regions other than sub-Saharan Africa, results largely depend on GCM scenarios and therefore are highly uncertain.

Fourth, although there is significant uncertainty on the effects of elevated CO_2 on crop yields, this uncertainty is carried to a much lesser extent on food security. This result emerges from a comparison of climate change simulations with and without CO_2 fertilization effects on crop yields. As can be seen, higher CO_2 fertilization would not greatly affect global projections of hunger. In view of the fact that essentially all SRES worlds are characterized by much higher real incomes, much improved transportation and communication options, and still sufficient global food production, the somewhat smaller supplies will not be able to make a dent in global food security outcomes. Many studies find that climate change without CO_2 fertilization would reduce the number of undernourished people by 2080 only by some 20 million to 140 million (120 million to 380 million for SRES A1, B1, and B2 without, compared with 100 million to 240 million with CO_2 fertilization effect). The exception again in these studies is SRES A2, under which the assumption of no CO_2 fertilization results in a projected range of 950 million to 1,300 million people undernourished in 2080, compared with 740 million to 850 million projected under climate change, but with CO_2 effects on crops.

Finally, recent research suggests large positive effects of climate stabilization for the agricultural sector. However, as the stabilizing effects of mitigation measures can take several decades to be realized from the moment of implementation, the benefits for crop production may be realized only in the second half of this century. Importantly, even in the presence of robust global long-term benefits, regional and temporal patterns of winners and losers that can be projected with current tools are highly uncertain and depend critically on the underlying GCM projections.

Uncertainties and Limitations

The finding that socio-economic development paths have an important bearing on future food security and that they are likely to top the effects of climate change should not, or at least not only, be interpreted as a probability-based forecast. This is because SRES scenarios offer a range of possible outcomes "without any sense of likelihood". Yet SRES scenarios, like all scenarios, do not overcome the inability to accurately project future changes in economic activity, emissions, and climate.

Second, the existing global assessments of climate change and food security have only been able to focus on the impacts on food availability and access to food, without quantification of the likely important climate change effects on food safety and vulnerability (stability). This means that these assessments neither include potential problems arising from additional impacts due to extreme events such as drought and floods nor do they quantify the potential impacts of changes in the prevalence of food-borne diseases (positive as well as negative) or the interaction of nutrition and health effects through changes in the proliferation of vector-borne diseases such as malaria. On the food availability side, they also exclude the impacts of a possible sea-level rise for agricultural production or those that are associated with possible reductions of marine or freshwater fish production.

Third, it is important to note that even in terms of food availability, all current assessments of world food supply have focused only on the impacts of mean climate change, *i.e.*, they have not considered the possibility of significant shifts in the frequency of extreme events on regional production potential, nor have they considered scenarios of abrupt climate or socio-economic change; any of these scenario variants is likely to increase the already negative projected impacts of climate change on world food supply. Models that take into account the specific biophysical, technological, and market responses necessary to simulate realistic adaptation to such events are not yet available.

Fourth, this review finds that recent global assessments of climate change and food security rest essentially on a single modeling framework, the IIASA system, which combines the FAO/IIASA AEZ model with various GCM models and the IIASA BLS system, or on close variants of the IIASA system. This has important implications for uncertainty, given that the robustness of all these assessments strongly depends on the performance of the underlying models. There is, therefore, a clear need for continued and enhanced validation efforts of both the agro-climatological and food trade tools developed at IIASA and widely used in the literature.

Finally, we note that assessments that do not only provide scenarios, but also attach probabilities for particular outcomes to come true could provide an important element for improved or at least better-informed policy decisions. A number of possibilities are offered to address the related modeling challenges. One option would be to produce such estimates with probability-based estimates of the (key) model parameters. Alternatively, the various scenarios could be constructed so that they reflect expert judgements on a particular issue. It would be desirable to attach probabilities to existing scenarios. Information on how likely the suggested outcomes are would contribute greatly to their usefulness for policy makers and help justify (or otherwise) policy measures to adapt to or mitigate the impacts of climate change on food security.

Climate change will affect all four dimensions of food security, namely food availability (*i.e.*, production and trade), access to food, stability of food supplies, and food utilization. The importance of the various dimensions and the overall impact of climate change on food security will differ across regions and over time and, most importantly, will depend on the overall socio-economic status that a country has accomplished as the effects of climate change set in.

Essentially all quantitative assessments show that climate change will adversely affect food security. Climate change will increase the dependency of developing countries on imports and accentuate existing focus of food insecurity on sub-Saharan Africa and to a lesser extent on South Asia. Within the developing world, the adverse impacts of climate change will fall disproportionately on the poor. Many quantitative assessments also show that the socio-economic environment in which climate change is likely to evolve is more important than the impacts that can be expected from the biophysical changes of climate change.

Less is known about the role of climate change for food stability and utilization, at least in quantitative terms. However, it is likely that differences in socio-economic development paths will also be the crucial determinant for food utilization in the long run and that they will be decisive for the ability to cope with problems of food instability, be they climate-related or caused by other factors.

Finally, all quantitative assessments we reviewed show that the first decades of the 21st century are expected to see low impacts of climate change, but also lower overall incomes and still a higher dependence on agriculture. During these first decades, the biophysical changes as such will be less pronounced but climate change will affect those particularly adversely that are still more dependent on agriculture and have lower overall incomes to cope with the impacts of climate change. By contrast, the second half of the century is expected to bring more severe biophysical impacts but also a greater ability to cope with them. The underlying assumption is that the general transition in the income formation away from agriculture toward nonagriculture will be successful.

How strong the impacts of climate change will be felt over all decades will crucially depend on the future policy environment for the poor. Freer trade can help to improve access to international supplies; investments in transportation and communication infrastructure will help provide secure and timely local deliveries; irrigation, a promotion of sustainable agricultural practices, and continued technological progress can play a crucial role in providing steady local and international supplies under climate change.

FOOD SECURITY, FOOD SYSTEMS AND THE LINK TO CLIMATE

There are several definitions of what constitutes food systems each formulated in relation to a specific range of issues. For GECAFSs purpose, food systems are defined as a set of dynamic interactions between and within the biogeophysical and human environments which result in the production, processing, distribution, preparation and consumption of food. They encompass components of: (i) food availability (with elements related to production, distribution and exchange); (ii) food access (with elements related to affordability, allocation and preference) and (iii) food utilization. Food systems, then, involve much broader considerations than productivity and production alone. They underpin food security, which is the state achieved when food systems operate such that 'all people, at all times, have physical and economic access to sufficient, safe and nutritious food to meet their dietary needs and food preferences for an active and healthy life'. Food security is diminished when food systems are stressed. This can be caused by a range of factors in addition to climate and other environmental changes (*e.g.* conflict, changes in international trade agreements and policies, HIV/AIDS) and may be particularly severe when these factors act in combination. Access to culturally acceptable food by individuals and communities, or means for its procurement, is increasingly being elaborated into human rights legislation.

Food systems may be simple, as in the case of a subsistence farmer who produces, processes and consumes food on farm. However, there are comparatively few individuals or households in the world that are totally self-reliant for food throughout a year, and in almost all cases there is an element of bartering, exchange, or the cash economy to bring food into the household. In many places, the food system has changed radically in the last century and continues to become increasingly complex. The intensification of agricultural production, since the 1940s has been accompanied by profound changes in the organization of food systems around the world including changes in distribution, marketing, affordability and preferences for particular food items. These changes are especially obvious in the USA and Europe, where market globalization has occurred with global sourcing of products by retailers direct from producers in the case of fresh fruit and vegetables and from a few, large manufacturers for other food products. Such changes have also shifted economic and political power from farmers to retailers, from national legislative bodies to regional and global organizations, and from the state to multinational corporations. For example, in a study of selected OECD countries in Europe, Grievink (2003) determined that the food chain has some 160 million consumers of whom about 3.2 million are also farmers or food producers, but the link between these groups is increasingly determined by the small number of food processors/manufacturers (about 90 000) and the even smaller number of buyers (about 100) for the supermarket chains.

Components of the food chain in selected OECD countries of Europe indicating the two inverted pyramid structure that relates farmers to consumers.

Food systems around the world are changing very rapidly as urbanization and globalization proceed apace. The urbanization of many predominantly rural countries in the last three decades has been accompanied by the rapid growth of supermarkets in many, often accompanied by foreign investment by global retail chains. However, even in poor countries such as Kenya, where per capita gross domestic product (GDP) was less than \$400 yr^{-1} in 2002, supermarkets have grown from a tiny niche market in 1997 to be greater than 20% of urban food retailing today. This growth has been a consequence of three major factors. First, rapid urbanization has seen the proportion of the population based in urban areas grow from 0.13 in 1975 to 0.36 in 2000 with an expectation of greater than 0.5 in 2013; second, trade and domestic market liberalization, since 1993 has seen the removal of import controls and the deregulation of prices and third, movement towards price-based competition between the indigenous chains. Success within Kenya is now spreading to other East African countries with important effects on the market conditions faced by farmers including the decline of traditional wholesalers (and the smallholder producers from whom they buy) and the increase in direct purchases from larger farms.

These marked changes in access and utilization of food around the world provide a context in which to evaluate the likely effects of climate and other environmental changes on crop production and food security. Not all food systems or parts of food systems are equally vulnerable to environmental changes because the capacity to cope with existing variability in bio-physical and socio-economic systems, and the ability of humans to perceive environmental changes and to adapt food systems, differs. Human vulnerability includes both the likelihood of exposure to stresses as well as the capacity to cope with such stresses (*i.e.* sensitivity). Vulnerability and poverty are often inter-related because both the likelihood of exposure to stresses is greater among the poor and because a large proportion of their resources are spent either purchasing or producing food, thereby reducing their capacity to cope with perturbations. Moreover, pursuit of food security frequently involves trade-offs with expenditure on health and education reducing still further the ability to improve longer-term living conditions or resilience to stress and shock. Food insecurity is experienced at a range of spatial scales from individual households to regions, as well as a range of time-scales, and reductions of vulnerability at one scale do not automatically flow to the next scale. For example, regional policy decisions do not always convert to successful local implementation especially if distribution services are inadequate, or food preferences are ignored.

Climate variation is one of several interacting factors that affect food security. For example, in studies of household food security in southern Africa, climate/environment was only one of some 33 drivers mentioned as important by householders. The mix of drivers varied across the region but in all communities many interacting factors resulted in vulnerability to food shortages. Overall, however, climate/environment was one of the seven factors influencing food security that were frequently cited, because of its role both as an ongoing issue (57% of cases) and as a 'shock' (43%). The impacts of sudden shocks such as drought are felt, then, on top of ongoing long-term stresses, and the low ability to cope with such shocks and to mitigate long-term stresses means that the employment of coping strategies that might be available to others, is at a too high cost or, simply, unavailable. Typically, reliance on purchased food increases in drought years due to losses in food production leading to an increase in poverty due to the synergistic action of other drivers such as rising food prices and unemployment. Scholes & Biggs (2004) record that the food security crisis in southern Africa in 2002–2003 was not simply a result of drought alone and, indeed, climatic stress was not as severe as in previous crises. Maize production during the preceding growing season was only 5.5% less than the previous five year average so food stocks at the start of the climatic shock were not unusually low. It was concluded that the crisis was indicative of entrenched vulnerability resulting from a range of regional and global political and economic factors including high food prices, legacies of structural adjustment, government policies, conflict and war, policies on genetically modified foods, and poor responses to the HIV/AIDS pandemic. The key here was that conditions, which weakened food systems in the region were already in place. The moderate climatic shock intensified food insecurity and the long-term vulnerability of the region.

The seven most frequently cited drivers in 49 studies of household-level food insecurity in southern Africa. The numbers in the arrows indicate the number of citations, as a percentage of 555 citations of 33 possible drivers. The drivers shaded in grey were noted as being chronic, while those in white indicate drivers experienced mainly as 'shocks'. The shaded arrows indicate drivers that acted primarily via reductions in food production, while the white arrows indicate those which acted by restricting access to food.

VULNERABILITY OF FOOD SYSTEMS TO CLIMATE CHANGE

Much climatic change–agricultural research has been focused on assessing the sensitivity of various attributes of crop systems (*e.g.* land suitability, crop yields, pest regimes) to specified changes in climate. These partial assessments most often consider climate change in isolation, focus on bio-physical aspects of production, and provide little insight into the food accessibility and food utilization dimensions of food security. To better address the food security concerns that are central to economic and sustainable development agendas, it is desirable to develop a broader research framework, which integrates bio-physical and socio-economic aspects of food systems and thereby addresses key questions including:

1. which aspects of food systems are most vulnerable to climate change? and
2. what can be done to reduce the vulnerability of these food systems and thereby improve food security?

The roots of vulnerability science can be traced back to famine and natural hazards but concepts developed in these areas have, to date, not been fully incorporated into climatic change/global environmental change studies. It has become clear from famine and hazards research that the key to assessing vulnerability is to develop research frameworks which can explicitly consider the social, economic and political constraints which condition the capacity of human systems (including food systems) to cope with external stressors such as climatic change, along with the magnitude and frequency of environmental stresses imposed on the system. For GECAFS, these concepts have been extended to consider environmental vulnerability (*i.e.* stresses originating from drought, storms and landslides and other such phenomena) and social vulnerability (*i.e.* the capacity of communities to cope with and recover from environmental stresses). The vulnerability of food systems is not determined by the nature and magnitude of environmental stress *per se*, but by the combination of the societal capacity to cope with, and/or recover from environmental change, coupled with the degree of exposure to stress. While the coping capacity and degree of exposure is related to environmental changes, they are both also related to changes in societal aspects such as institutions and resource accessibility. Finally, changes in the food system aimed at reducing vulnerability feed back to environmental and societal changes themselves. They may, for example, reinforce agricultural practices that either reduce or exacerbate land degradation, and increase or reduce farm profitability.

Initial research in the Indo-Gangetic Plain (IGP) has used this more integrated approach to help define the vulnerability of the region's food systems. It has demonstrated that the conditions underpinning vulnerability are not uniform throughout the region. In the western IGP (a region of general surplus production), food systems are most vulnerable to issues related to the availability of water; excessive irrigation has lead to rising watertables and soil salinization in some areas while in others water shortage has resulted in falling watertables, rapidly increasing costs of pumping and shortage of drinking water. In the eastern IGP, resource poor farmers, who have very limited options to cope with and recover from external stresses, are most vulnerable to environmental changes such as rising sea-level, and climate change and climate variability leading to increased risk of flooding. Overall, this analysis suggests that food insecurity concerns cannot be effectively addressed by a single region-wide policy.

ADAPTATION: REDUCING THE VULNERABILITY OF FOOD SYSTEMS TO CLIMATE CHANGE

Development of human societies has involved a continuous process of adapting to changing stresses and opportunities. While climate change is seen as a relatively recent phenomenon, individuals and societies are used to adapting to a range of environmental and socio-economic stresses. In many parts of the world, and especially in semi-arid lands, there is an accumulated experience with phenomenon such as drought. As climate extremes are predicted to increase in frequency and intensity in future, it is important to understand and learn from relevant past adaptations and indigenous knowledge. However, changes in climate variability and mean values will bring additional complications to many, especially those dependent on food systems that are particularly vulnerable to these additional stresses. Food systems fail to deliver food security when related determinants, and/or the links between them, are disrupted by climate change or other stresses. The food systems approach to research on food security allows for adaptation options aimed at reducing vulnerability to be considered in terms of any of these determinants and the stresses acting upon them. Adaptations may occur in relation to, for instance, agronomic or fisheries aspects regarding food production; or government-set prices and incomes concerning access to food; or changes in societal values concerning food utilization. The key issues for adapting food systems to reduce their vulnerability to climate change are to:

1. identify which related determinants are particularly sensitive to GEC;
2. enhance effective related determinants; and
3. restore disrupted related determinants.

To translate these theoretical considerations into practice requires research to:

1. identify and evaluate the possible adaptations mechanisms to reduce food system vulnerability to climate change;
2. identify, document and learn from past and current coping mechanisms employed by vulnerable groups in their day-to-day food supply systems;
3. analyse and strengthen the capability of communities and countries to adapt as much as possible; and
4. identify the most suitable level at which each adaptation strategy should be implemented.

These points are illustrated below using an example from each of the three elements of the food system.

Reducing Food System Uulnerability

Past increases in agricultural production have occurred as a result of both extensification (altering natural ecosystems to generate products) and intensification (producing more of the desired products per unit area of land already used for agriculture; Gregory & Ingram 2000). In future, intensification will be the dominant means for increasing production although the cultivation of new land will be important in some regions. Increased yields per unit area, with a smaller contribution from an increased number of crops grown in a seasonal cycle, is expected to be the main way in which crop production will rise to meet demand. In the recent past, such increases have been achieved by a 'unique conjunction of three innovations', namely cheap nitrogenous fertilizers combined with semi-dwarf genotypes of cereals, effective weed control with herbicides, and the expansion of irrigation. For the future, continued technological developments are anticipated to facilitate the adaptation of crops to changing environments.

Porter & Semenov (2005) has described some of the plant traits that may permit adaptation to climate change. An area in which progress may be possible is in improved selection of genotypes that utilize limited supplies of water stored in soils; less rainfall may be a consequence of climate change or increased climate variability in some regions. There are many plant characters and elements of crop management that contribute to the efficient use of water by crops (Gregory 2004), but relatively little attention has been paid to root characters that may allow more water to be exploited or used more efficiently, largely because root systems are very difficult to measure. However, genotypic differences are known to exist in many features of root systems (*e.g.* depth of rooting, rate of downward extension, diameter of roots, total length) which may be exploitable to improve crop yield in drier climates. Studies with existing genotypes in dry areas may inform the adaptation possible under conditions of changed climate. For example, in the Mediterranean environment of northern Syria, crops are largely dependent on the use of growing season rainfall (very little water is stored from season to season). Studies of the root growth, water use and yield of the local landrace (Arabic abiad) at sites with typically less than 350 mm annual rainfall, showed that this genotype consistently had greater root lengths per unit soil volume at depths below 15 cm than other genotypes such as the variety Beecher, and that this was associated with greater water uptake in the 3–4 weeks before anthesis. In this environment, faster rates of growth before anthesis were reflected in higher crop yields; senescence of the crop before the grains could fill ('haying off') was not observed. The development of DNA-based molecular markers has opened up opportunities for identifying the genetic factors (quantitative trait loci) underpinning various root traits. Again, this science is at an early stage of development for root traits, but significant progress has been made in studies of drought tolerance with rice.

The distribution of root length with soil depth for the barley genotypes Arabic abiad (white bar) and Beecher (black bar). The distributions are the average for two sites in northern Syria with contrasting soils and rainfall .

Improving Food Distribution

Infrastructural and non-infrastructural controls on food distribution can be significant impediments to reducing food system vulnerability in a timely manner. This became strikingly apparent during the drought relief effort mounted in 1990/1991 in southern Africa in response to the estimated 86 million people at risk in the region (of whom some 20 millions were deemed at 'serious risk').

A massive international food aid programme was launched with food to be delivered via a number of rail 'corridors' from the region's major ports to the hinterland. Nearly, 8 million tonnes of food grains were imported by the relief programme for the 10 countries affected; this was almost four times the normal annual rate of imports for the region. This substantial increase in imports resulted in pressure on the region's distribution systems, leading to a number of problems which would not have been significant in 'normal' years. These are well illustrated by examples taken from the Maputo and Beira corridor. Infrastructural constraints included ongoing rehabilitation, physical impediments (steep gradients and tight curves at which points grain was stolen from the slow-moving wagons), a shortage of bags and tarpaulins, the need to tranship Malawi-bound cargo in Harare, and a shortage of rolling stock and locomotive power. In addition, a number of non-infrastructural constraints further complicated the situation. These included general security problems along the corridor, regulatory constraints for cargo destined for Zambia, conflict between humanitarian requirements and commercial concerns, poor labour management systems in ports (where there were no incentives to work more than necessary), and transit toll fees in Mozambique.

Food availability for the region was severely constrained due not to lack of food *per se* (there were ships queuing at anchor to unload), but by lack of investment in distribution systems and institutional constraints. This brief summary highlights several ways in which regional food insecurity could be reduced, and shows that adaptation options can include a range of issues including, among others, regional investment in port, rail and grain storage infrastructure and in region-wide political agreements to facilitate the flow of food in an emergency.

Increasing Economic Access to Food

Improved economic access to food is an important development goal, but the means of achieving it and the consequences of strategies aimed at its achievement are the subject of much discussion. In the case of southern Africa, Arntzen *et al*. (2004) indicate that the discussion centres around varied means. First, price mechanisms and policies could be designed that serve the interest of producers (incentive to produce more food) and consumers (to facilitate access to food). Second, regional specialization in food production and regional trade would lower production costs and food prices and, therefore, improve access. This important adaptation is as yet hardly pursued, but should gain momentum with trade liberalization and policy shifts towards food security. Third, economic growth will lead to income and employment generation, both of which will facilitate access to food. Finally, stability and governance supported by an effective pool of human and institutional resources facilitate the establishment and maintenance of food systems.

In south Asia, studies have examined how income growth has led to changes in diets away from traditional foods. This may have negative impacts on local farmers who grow traditional foods and are not well integrated into markets. The impact of trade liberalization on the poor is a topic of current study, but there is an emerging consensus that they should be protected from negative impacts through the implementation of safety nets.

6

Fluid Boundaries of Indian Forestry

INTRODUCTION

The community forest management on (state-owned) forest lands in India. It examines the policy framework of Joint Forest Management (JFM) that has, in a radical breakaway from earlier policies that keeled on centralized, revenue orientated control of forests, promised several managerial and usufruct concessions to forest citizens. The paper endeavours to answer the extent to which decentralized forest management can be fostered in varying political environments with marked contrast in social capitals within the stakeholders. It compares forestry practices amongst the members of Forest Protection Committees (FPCs), an institution formed on JFM initiative, in two provinces in eastern India, Bihar and West Bengal. The paper examines, both formal and informal, functionalities that have found their ways into JFM-prompted forest management. Using a political ecology approach, the paper, examines features that mark out forestry practices at village level and recommends adjustments in institutional design of JFM, as well as in nature of interception that are made by various agencies (NGOs, *panchayats*) involved with rural development forestry.

JOINT FOREST MANAGEMENT RESOLUTION

Joint Forest Management resolution follows a history of contest between forest dependent user groups and the state. It itself, however, encompasses, in rhetoric at least, themes of participation, equity, decentralization and grassroot democracy. The impact of this new paradigm has been evident in several ways. Forest protection committees, as the paper demonstrates, are playing active roles in negotiating the transaction of forest usufruct within and outside their communities. This includes approval of indigenous extractive methods of a community and fulfillment of various forest-based needs, and even creation of 'zones of exclusion' where village forests become out of bounds for outsiders. Forest Protection Committees (FPCs) also assert themselves against the traditional village institution, the *panchayat* or the village council. While the Forest Department's role in 'policing and fencing off' of forest has somewhat receded, they have not met villagers' expectations of camaraderie and support. Instead, villagers carry *personalized* notions of participation, while managing forest in spirit of a 'common' property resource. The paper, however, argues that one cannot undermine the constraints that an over specialized institutions like FPCs face. At village level, a rigid dichotomization cannot be held between forestry and other rural development issues. It would be imperative, for an institution such as an FPC to address larger issues on social, educational and development lines in order to gain sustainability and legitimacy.

The first part of the paper introduces key political ecology themes and schools of thought vis-à-vis forest management in India. The recent participatory forest schemes are then discussed. The sections that follow discuss case studies, where Forest Protection Committees (FPCs) in states of Bihar and Bengal are examined to address areas that might reasonably be supposed to give life to village-based ecological institutions.

RESEARCH OF THIRD WORLD POLITICAL ECOLOGY

In the last two decades, research has examined extensively, if somewhat unevenly, the contextual sources of environmental change. Scholars from diverse academic and institutional backgrounds have examined links between environmental and political activities in parts of Asia, Africa and Latin America. The studies have further been combined with cultural-social interpretations. Correspondingly, a body of work that may be termed Third-World political ecology has emerged.

A great part of research by political ecologists has focused on forest resources. These studies are often nation/ area specific and have tended to examine state as against the actual user groups. The work includes that of Shepherd (1985, 1986, 1988, 1989), Nesmith (1991), Guha and Gadgil (1989, 1993, 1995), Vira (1995), Corbridge and Jewitt (1997), Mawdsley (1997), Muldavin (1996), and Nanang and Inoue (2000). In the process, many schools of thoughts have got established, and where, also, many stereotype views have been challenged. The concept of sustainable indigenous practices (as against scientific forestry) has been pushed forward, villagers' priorities and problems have been elicited, and a gendered approach to rural environment and economic opportunities has been made. A common theme in many of the theses is the struggle that exists between the state and its people over control, use and management of forest land. Another definitive element has been the increasing concessions forest citizens have come to receive from the state over control of forest land, and the ways such participation have been institutionalized. This phenomenon has been documented extensively: greater participation of the peasants, in 'farmers' first' model, has been elicited, role of women has been sought , existence of heterogeneity in rural societies has been appreciated and adjustments made by the Forest Department has both been commended and criticized.

It examines community-access to forest resources in Bihar and Bengal and argues that despite the state's secure tenurial rights over forest land, the JFM resolution has given the populace, the tools they have begun to use not only for forestry but also for larger issues of livelihood and development. The general theoretical and policy context of this research. The first part reviews the debates and literature on history of forest management in India. The participatory phase of Indian forestry that started taking shape from 1970s onwards. The case studies, that follow, further illustrate the points made in these sections.

Forestry in India: A Brief History

In its bid to 'manage' forest, the state often found itself in conflict with the forest-dependent rural communities. The ecological history debates have addressed questions as to which phase, the British (and post-independence) or pre-British, of forest management was more pertinent & just, and have brought available records under close scrutiny. However, arguments and conclusions vary greatly. A school of thought led by Ramchandra Guha has challenged the central premises of the imperial historians - that the colonisers saved the forests of South Asia from certain destruction by indigenous forest users. Guha proposed that during the mid 19th century three schools of thoughts had developed with regard to the future treatment of forests and forest land:

"the first, that of annexationist held out for nothing less that total state control over the forest areas. The second, that of he pragmatics, argued in favour of state management of ecologically sensitive and strategically valuable forests, allowing other areas to remain under communal systems of management. The third position (a mirror image of the first), the populists completely rejected state intervention, holding that tribals and peasants must exercise sovereign rights over woodland. (However), of the three, the annexationists triumphed".

Forest-land Acquisitions

Indeed, the corollary to Guha's annexationist theory is that such forest-land acquisitions and alienating land-tenure arrangement give forest-dependent communities a moral right to claim back their 'share' in natural resources. Guha and (Madhav) Gadgil's idea of the predatory state finds support in select statements of colonial foresters. Ribbentrop, for example, argued that scientific forestry under imperial aegis marked the end of a 'war on the forests'. Stebbing argued that rapacious private interests were brought under scientific supervision and control in the colonial period. Guha in turn has held that the practices of colonial forestry were largely an outgrowth of the strategic and revenue needs of empire. Gadgil suggests that the period up to 1800 A.D., was a time of 'equilibrium' between people and nature. The detractors of Guha & Gadgil, on the other hand, have tried to produce evidence from ancient Indian literature, however much incomplete, that forest practices could have been as harmful and imperialist in approach as in British India.

History of Forestry in Ancient and Medieval India

Records on control and management systems of forestry in ancient and medieval India are far from complete. Information about forests during this period is available from dotted sources such as Kautilya's *Arthashashtra* (321 BC), *Indika* by Megasthenes (305 BC), the inscriptions of Ashoka (273 BC to 236 BC), *Akbar-nama* (1650 AD) and so on. The dominant theory accounting for forestry practices in this period argues that the Aryans who migrated into India during the second millennium BC fought with the indigenous population (now represented by the tribal-groups) then 'living in the forest', driving them to remote areas. Recent appraisal h as, however, presented alternative views. Das, for example, contests the version of tribals being driven to forests and maintains that they were essentially agriculturists.. The administrative category 'tribe' given by the colonial administrat on has obscured the fact that forest dwellers have been peasants for a long time

The Aryans cleared the forests with their superior iron tools, and started organized agricultural and forestry practices. During the reign of Chandra Gupta Maurya 320 B.C., there was a regular forest department administered by the *Kupyadhyaksha* (superintendent of forest), assisted by a number of *Vanapals* or the forest guards. Forests in the Mauryan Empire came under direct control of the sovereign, and both ownership and control on use was highly differentiated. Forests remained a major source of revenue in Gupta period (320 - 550 A.D.). There are records from the period that indicate that there was a collection of revenue for forest-use from the public.

DOCUMENTATION ON NATURE OF FOREST MANAGEMENT

There is little documentation on nature of forest management in medieval period inscriptions, though the inferences are that the *Mughal*-rule (1400 A.D.-) saw forest resources as *pleasure* objects. Extensive hunting (for sport) is recorded during the period. Although rulers, particularly Jehangir and Akbar, were fond of roadside trees and gardens, they are argued to have shown little interest in forest conservation, and extensive damage was done during wars with other ruling states.

The reason to sketch out the countenance of ancient and medieval forestry is to emphasize that forest practices could have been as harmful and imperialist in approach as in British-ruled India. More specifically, it seeks to indicate that the state, once finding a forest resource of value, has, through time, tried to exercise control over it. Secondly, as different groups contest for (albeit with different incentives) control, a hierarchy comes to exist both in use and in perception of forests. Hence, where state, for example, could pursue territorial and economic interests in forest terrains, the forest citizen utilized their immediacy to evolve their own set of practices to secure forest product for subsistence and household economy. The various 'niches' in this hierarchy is not always exclusive of each other, and indeed has resulted in the oft-cited contests between the state and the forest citizens. The meaning of participation, hence, would best be found in use of forest in manners that can serve to fulfill needs of a multiple set of users situated at various levels.

Colonial Forestry (1800 – 1947) in British Administration

The newly established British administration in India was initially not alive to the need for careful husbanding of forest resources, and was under the impression that the forest wealth of India was inexhaustible. The British themselves were new to the ideas of scientific forestry, and had no developed forest organization in Britain. The first step in British Indian Forestry came in South India. In 1800, a commission was appointed to enquire into availability of teak in the Malabar Forests. In 1805, a Forest Committee was constituted to investigate the capacity of forests and the status of proprietary rights over them. Then in 1806, teak (Tectona *grandis*) was reserved as a royal right in parts of south India. Later, in 1855, Governor-general Dalhousie promulgated for the first time an outline for forest conservancy for the whole country called the 'Charter of the Indian Forests'. Following this, Deitrich Brandis was appointed Superintendent of Forests in India in 1856. The Forest Department under him proceeded to transform the working of India's forests, from the initial practice of exploiting them to obtain supplies of timber, to treating them as a growing biological entity of much value, and handling them in accordance with the principles of scientific forestry.

In 1857, after the Indian mutiny, India came under the direct rule of Britain. It was a turning point not only for the civil administration of the sub-continent but also for forest regimes. India was now to be ruled not only for trade interests, but also was taken over by a nation, which was obliged to have a greater control over the territories of her colony. In his letter, dated 1st November 1864 to the Secretary of State for India, the Governor-General pointed out that the idea of allowing individual proprietary rights in forests must be abolished, as such rights might lead to the destruction of forests. However, local governments were allowed to control forest management and were to be only given with policy guidelines by the Government of India. The first major Forest Act came into being in 1865. Under this Act, local governments were empowered to draft rules for law-enforcement in their respective regions. A revised Forest Act in 1878 provided for constitution of Reserved and Protected category forests. During this period (1880-1900) forest *settlement*, demarcation and survey of forest land were actively in progress in various provinces. Later, following Dr. Voelcker's report (1897), the Government of India declared in its forest policy, that permanent cultivation should come before forestry, that the satisfaction of the needs of the local population at non-competitive rates, if not free, should over-ride all considerations of revenue, and that after the fulfillment of the above conditions, the realization of maximum revenue should be the guiding factor. Rapid progress was made in organized forestry in the years between 1925-47. There was a long history of working plans and forest-research spanning more than eighty years of 'scientific' management of Indian forests by the time India gained independence in 1947.

Forest Management since Independence

The Indian Forest Act of 1927, which is modeled on the earlier act of 1878, still defines the legal framework for forest management in the country. Current issues of biodiversity, equity and customary rights do not find prominence in the legislation. With the abolition of *Zamindari* in 1951 large tracts of private forest land were vested with the State, making the Forest Department, as Guha says, the "biggest landlord in the country". The National Forest Policy of 1952 was strongly hinged on scientific forest management which drew largely from existing forestry practices in USSR and the USA. The practice was to encourage maximization of forest revenues and to give priority to agriculture and industry. The second five-year plan (1952-57), which had followed the National Forest Policy, however, also directed the States to aim to put 33% of their land under forest cover.

INDIGENOUS RIGHTS AND ACCESS TO FORESTS

It is apparent that in the nineteenth century and right up to the 1970s, indigenous rights and access to forests were contested by state agencies claiming to act in the national interest. Many forest-dependent communities, tribal (*adivasi*) groups in particular, were stigmatized as forest-destroying communities. In the face of these contests (and rapid degradation of forest cover) there has been growing support in India for the idea of placing some 'power' back in the hands of the rural poor. Since the late 1970s a number of social forestry programmes have been developed and the emphasis of state forest policies has shifted from commercial forestry to that of meeting needs of the forest dependent population on a priority basis. The most recent of these programmes (1988 -) has embraced a philosophy of Joint Forest Management (JFM).

Social Forestry Development

Following the Agricultural Commissions' recommendations in 1976, forests that had remained under total control of the states became a *concurrent* subject. The government in New Delhi was to play central role in formulating policy guidelines and in coordinating forestry programmes. As the interests of the centre and the states do not necessarily coincide, the government in New Delhi could arguably take a long-term view. This arrangement, allows conservationists, international agencies and voluntary groups, despite their low numerical strength, to have considerable influence on the direction of policy at the level of the centre. Also, state governments are subject to regional and local pressures, which may not be a factor for decision making at the centre, it was argued.

Raise Forest on Private Land

In the 1970s members of the public were invited to raise forest on private land. By that time it was widely recognized that destruction of forests would not cease without obtaining participation from forest dependent communities. The nation had by then geared to address the subsistence and cultural dependence of its people. The period coincided with the social justice call of Prime minister, Indira Gandhi (5th five year plan). Reflecting the national rhetoric of social justice and equity, rural communities were to find increased (official) access to forest usufruct. Thus a shift, from pure regimes to shared regimes took place. India as a leader in social forestry) took a series of participatory interventions: village woodlots, strip plantation (roadsides, alongside railways), farm forestry, agro-forestry, home gardens, urban forestry, fuelwood plantations, festival of trees (*van mahotsava*), education camps, nurseries, and research & training.

However, a clear disjunction between the programmes' intended goals and actual products emerged. While the targets of farm forestry (in many states), for example, were considerably over-achieved, those from community forestry were less successful. It were larger farmers who primarily took up farm forestry and adopted commercially viable species, such as eucalyptus , for plantation. Villagers invariably had an incentive in income generation. Also, the project was alleged to have encouraged little participation from tribal population and villagers were reluctant to make long-term investments (Shepherd, 1985). Vandana Shiva, the eco-feminist writer, has been vociferous against the programmes which she thought promoted mono-culture , did not encourage involvement from women and from which particularly poor seasonal migrant labourers lost employment as tree planting is not a labour intensive occupation. Afforestation of wasteland to her, hence, was 'privatization' of the commons. Social forestry found willing funding bodies in World Bank, USAID, DANIDA, ODA, SIDA, EEC, OECF. These donors progressively attempted to influence the direction of policy especially in regeneration of degraded forest areas and for wildlife conservation. The international agencies came under attack for promoting a 'top-down' approach which the participatory programme was to do away with. They were alleged to have vested interest in country's corridors of power . Also, the forest bureaucracy was criticized for its inability to shake off power structures that are conditioned to emphasize policing, protection, and state production. Also, the department represented a long history of protecting the interests of forest-dependent industries. These factors despite policies at the centre to give priority to fuelwood and subsistence needs of the village community, failed to marginalize the private commercial interest in the forests.

Joint Forest Management: The Current Paradigm

Notwithstanding the criticisms and failures that social forestry faced, the programmes succeeded in initiating the processes for participatory forest regimes, devolution of decisions and recognition of villagers' subsistence needs. In 1990, the Ministry of Forestry and Environment, in New Delhi advised the state governments to take up Joint Forest Management (JFM).

Not surprisingly, Joint Forest Management systems could get attention both from the group (like Guha and Gadgil) that wishes to see in JFM, the chance of restoration of moral economy through increased tenurial and usufruct rights for the rural communities and exercise of 'sustainable' indigenous practices, and those (like Rangan) who, on the other hand, are keen to find solutions within existing institutions by way of 'pressuring states to intervene on behalf of marginalized communities, to ensure equitable access to the potential benefits of economic development'. Results from JFM have indeed been encouraging. JFM is reported to be instilling new attitudes and behaviour in forest bureaucracy towards the villagers , and vice versa. Many forest beat officers have reported that their primary incentive to encourage JFM is the vastly improved relations they have come to cultivate with the community. NGOs too have found themselves accommodated in the new regime , as an agent of change and promotion. Through group exercises, discussions, meetings and monitoring they are to instill participatory practices both in members of Forest Protection Committees and the Forest Department officials.

There have been several shifts in management of forests by the state and these, in turn, have been reflected in attitudes adopted by the user groups. The people, in the current setup, have come to view the state, NGOs and grassroot organizations (*panchayats*, for example) as bodies they can use to bring development in their community. The solution to equitable use of forests is not to isolate forest citizens in realm of some imagined sustainable practices but to chart out on what sorts of institutional links and relationships could be established between the rural communities and other institutions to enrich this dialogue.

The chapter contends that while many predicaments arising from forest conflicts between state and people have been successfully attempted at, the state not only has to take a more imaginative approach to include the practices that flourish informally at FPC level, but also reach out to intervene in various facets of rural development forestry. The findings here constantly are pointers to the peoples' desire to involve the state (and other institutions) for development and access to market and available funds, as against leaving out the state in such processes.

Research Methodology

The paper compares the States of Bihar and West Bengal, sharing contiguous boundary and similar socio-cultural environment and forest ecology. It examines the characteristics that mark forest activities (extraction, consumption, protection) among forest citizens in Bihar and Bengal. The hypotheses that are tested include that the Forest Protection Committee, a key stakeholder in JFM, will be motivated to invest time and other household resources to meet the *mandatory* demands of JFM (forest-guard duties, monitoring, attendance in meetings, forest works, etc.) over a number of years in anticipation of sharing the usufruct gains associated with afforestation and other forest management schemes; that the age old conflict over establishment of rights over forest resources would be amicably resolved (now that government recognizes the claims of villagers, the villagers would in turn recognize the jurisdiction of forest land boundaries); and finally, that the 'equity' rhetoric would be served. JFM would aid the local community in various village development works, and in increasing their income.

The means we have depended on include literature-review and archival analysis. Theories and archives provide the background to which case-studies are directed, and against which the results from the case-studies have been tested. The research looks into positions taken by ecological institutions in the JFM set up. The field work included interviews, group discussion, attending meetings, participation-observation, observational techniques and secondary data collection (official documents/ progress reports). Interview methods were mostly semi-structured and open-ended. Research carried out in one State was mirrored in another.

JOINT FOREST MANAGEMENT IN BIHAR AND WEST BENGAL

The theoretical premises that have been established earlier. This is achieved by providing a geographical setting to the research and by discussing case studies from Bihar and West Bengal. After a brief note on the institutional history in these two States, the section spells out formal aspects of the Joint Forest Management systems in each of the provinces, and, then, examines the activities that mark the actual boundaries to perceptions and practices in local forests.

West Bengal and Bihar: At Odds on Village-Level Governance

During most of the British rule, Bihar remained a part of the Bengal Presidency and, until the partition of Bengal in 1912, was governed from

Calcutta. In 1912, Bihar and Orissa together formed a separate Governor's province. In 1936, prompted by the Re-organization of State Act, the five British districts of Orissa formed still another province. In 1947, after Independence, Bengal was further partitioned into east and west, the former becoming a part of the newly formed Pakistan.

The left coalition Government of West Bengal has been progressive and committed to land reforms and other social programmes to empower poor rural communities. Starting in 1977, share croppers and land-less persons were given 99-year lease (*pattas*) on lands that were vested with the State, soon after independence, after the Land Ceiling Act. In 1979, the State Government vigorously implemented 'Operation *Barga*'. These programmes resulted in provision of land to the landless, and tenurial security to share-croppers and small land holders. Also, *Panchayati-Raj* (village level governance) was strengthened during the period. The members of *Panchayat* were encouraged to participate in decision-making and overall management of rural development.

Rural Infrastructure

The strings of achievements that Bengal attained within rural infrastructure not only instilled confidence in the bureaucratic set-up in general, but also complemented grassroots level institutions which could then deal effectively with the initiatives being provided by the foresters. The Forest Department to its credit took many innovative steps. West Bengal Forest Department issued a number of facilitating resolutions to foster participatory environment. In 1980, the State Government's '*New* Directives on Forest Management' gave privileges and concessions to tribal population living in vicinity of forests. Later, complementing the success of Arabari experiment, forest department issued orders in 1987 to share 25 per cent of the net return on harvest of Sal poles with the participating communities. Also, a new scheme titled 'Economic Rehabilitation of Fringe Population' was drawn up through which various development schemes were taken up in the forest fringes to generate income for people through forestry based activities. Other socially beneficial measures such as drinking water supply, construction of earthen dams and minor irrigation projects were planned.

In Bihar, by contrast, the lands that were vested with the State could not effectively be re-distributed to the landless. *Panchayat*s were often dominated by prominent villagers from the caste-Hindu population. The introduction of *Panchayati-Raj* (made redundant in Bihar now, by the High Court, because of over-due elections), also coincided with breakdown of many traditional systems of administration among the tribals. The Forest Department, responding to participatory rhetoric in the nation, started Social Forestry projects in the 1980s. The programme was 'in operation' until 1990. The results, however, were unconvincing. The survival rate of the farm forestry plantations (funded by Swedish International Development Agency (SIDA)), for example, was as low as 20 per cent. Joint Forest Management was officially announced in 1990 in Bihar. The forest bureaucracy, however, has been unable to shake off the power structure that are conditioned to emphasize policing and protection. The department is further plagued by both structural constraints, like insufficient number of staff, lack of funds, and constitutional bottle-necks (for example, the working plans that enable harvest of forests have not been sanctioned in the State for past several years).

POLICY IMPLEMENTATION AND INSTITUTIONAL HISTORY

Despite a strong contrast, particularly in policy implementation and institutional history, many commonalties are found in use and perception of forests at community level in both Bihar and Bengal. The framework of JFM, in its institutional design, for example, is similar in both the states. Also, while the state has eased its grip over forests, it has not let the tenurial-right over the forest land slip and almost all management decisions, including formation and annulment of committees, are taken by the Forest Department. The following sections test a number of obligations, prescribed as part of Joint Forest Management, to see how participatory rhetoric has filtered through and found place among forest citizens.

Setting of the Study-Areas

The southern part of Bihar and the adjoining south-western parts of Bengal not only are more forested (with high community dependence) but also contain the largest extent of protected-type of State forests, where Joint Forest Management arrangement is, principally, sought to be instituted. Also, the population make up in south Bihar and south-west Bengal is mix of caste Hindu and tribals with pockets of Muslim residents. The case study in West Bengal refers to Gopegarh Forest *Beat* in Midnapore East Forest Division, and mainly, to Bhagwati Chowk village. Chowk is a small village with 37 households made-up of a near uniform population and is next to a relatively large patch of forest. In Bihar, the research was carried out in Annagarha *Beat*, in Ranchi East Forest Division. The paper focuses mainly on Ober village (with 200 acres of forest land and as many households). In both the areas, tropical deciduous forest forms the major forest type, with Sal (Shorea *robusta*) as the dominant tree species. Some of the area, however, have extensive Eucalyptus plantation, a reminder of social forestry plantation schemes.

SOCIAL CAPITAL: LEADERSHIP, RECIPROCITY, AND INSTITUTIONAL SUPPORT

The section locates the two regions in the wider socio-political setting of their respective States. It examines issues on dependency on forest, leadership, and involvement of other institution such as NGO and *panchayats*. It is argued that while Forest Committees have been able to assert themselves in presence of other institutions, they, as over specialized bodies, often lack the sturdiness to withstand influences (often, detrimental) from these bodies.

The Forestry Case of Bihar

The forests in Annagarha had belonged to the landlords and were vested in the State soon after the Indian independence. The landlord was from the tribal Munda community from the village of Nawagarh. After independence, he was elected to head the *panchayat* in the area. Being the centre of local governance, the Munda community of Nawagarh has been able to have access to development works (roads, culvert, tube wells, school, for example) and institutional support (police, post-office, information network) more effectively than the neighbouring Ober village. The Bediyas that make up Ober village (which is within the jurisdiction of Nawagarh *panchayat)*, on the other hand, have often found themselves in isolation when it has come to their relationship with state's machinery. Vis-à-vis the forest, their memory is of progressive assertion of rights by the state, monopolizing the forest land and its resources. The Bediyas of Ober are principally agricultulartists with small per capita land holdings.

The dependency on forests is in vital areas such as fuelwood, poles of Sal trees (for construction of houses and ploughs), and also as a source of food in drought conditions. Resultantly, they have as a matter of routine taken out usufruct from the adjacent forest and remained in contest with the patrolling staff of the Forest Department. The participatory paradigm, particularly after the JFM, which asks of the villages to form committees have resulted in forums from which individuals have tried to legitimize indigenous perception of rights over the forests and their own forest practices.

Community's Religious Headman

The residents of Ober were guided by the community's religious headman, Puran Mahto. This traditional leadership was, however, challenged by a resident of the *neeche toli* (the part of village that lies in a basin), Lahru Bediya who, since 1976, involved himself in 'social service'. A secondary school pass out, he left a lower rank government job, to work for village welfare. Interestingly, forest was an issue he addressed from the beginning. The year 1976-77 was a time when mass scale forest felling was taken up by the State. This was often carried out with the help of private contractors. The period exposed the villagers to commercial value of timber, and eroded the 'logic of conservation' which Forest Department had until now advocated to these people. They too got into the practice of an indiscriminate felling and would often collaborate with contractors. Forest had acquired an open access resource image: one which could provide the people with a source of quick revenue with little accountability.

Delineate Forest Boundaries

It was during this open-access attitude, that Lahru Bediya closed ranks with like minded people and delineate forest boundaries that would be out of bounds for outsiders – in effect, marking out an area in forest, called as the *rakhat* (or the protected land), which would be under the management of the residents of Ober. The first major fight that Lahru Bediya took up with the outsiders was in 1981.

The villagers from the neighbouring Bediya village of Sringeri would regularly fell timber in the forests of Ober *khas toli*. The fight that he took up was, although against the wishes of the headman Mahto, resolved in his favour both by the government official at Sringeri and the Munda headman at the Nawagarh village-council.

By 1984, Lahru had demarcated the forest boundary of Ober and initiated the practice of keeping a log-book that would record minutes from discussion during various meetings. The records demonstrate that the new forum evolved with time to make decisions on issues ranging from getting consensus within the village on benefits of the forests and need to protect *rakhat* forest, to one that would stop and fine the outsiders who enter into their forests Additionally, there is a general respect for the forest patch next to an individuals agricultural land, called jote. The forest land next to jote is protected by the owner of that field and is usually kept as a resource that could be used at a time when forest usufruct is unavailable from elesewehre or from the rakhat, the community forest.

With increased acceptance in the community the forum became active in other social areas as well. Within a year's time the meetings would decide on social and moral issues of the Ober *khas* community. By 1992, Lahru had challenged the leadership of the religious head man, Puran Mahto on grounds of Mahto's inability in bringing in development in the village, he proposed the name of Balsu Bediya as the new Mahto of the village. There are evidences that suggest that successes (and failures) of Lahru's initiatives have spatial connotations. First, it was easy to organize support on the defense of a forest that was in immediate environments. Second, he could provide in himself a proximate alternative to leadership - the other alternative being the *panchayat* members of the Nawagarh village. Third, the organization of the support that Lahru enjoyed shows a spatial pattern. Lahru, a resident of *neeche toli* always enjoyed a loyal following from his own *toli*. The Mahto leadership, on the other hand, in *uppar toli* stood strong against Lahru's resistance, and it took 3 years for the new Mahto, who is from Lahru's *toli*, to replace the old one.

Control over Forests

As the community wrested out the control over forests from the neighbouring villages, Lahru provided a leadership that could not only ensure equitable access to forest but also pull in various aspects of livelihood and development on the forest platform. Indeed, the latter would prove to be a keener incentive for the community to join hands. The vulnerability of the leadership of Lahru was exposed when, in October 1987, Lahru Bediya was away for a week to a nearby town, he was to find, on his return, that most of the community *rakhat*-forest has been cut down by the villagers.

In his absence the residents could neither resist the outsiders nor their own temptation to strike down the forest for quick revenue. They after a gap of five years of protection wanted returns of their labour and patience. They viewed Lahru's attitude towards the forest as autocratic and one that was dictated by personal agenda. Lahru Bediya had little option but to lay dormant for two years. In the same year, an NGO called Ram Krishna Mission, which had been operating in a few nearby villages on rural development, got involved with Ober khas toli. This external force with more potent incentives was to challenge and marginalize Lahru's leadership even more.

The NGO is involved on issues of tribal development mainly through mobilization of government and private funds, and institutionalizing norms like training, meetings, gift of labour for development schemes, pooling in funds etc. in the villages.

In Ober it has opened a clubhouse, and incidentally taken two youths to head the organizations' activities in the village. Though the NGO-sponsored forum primarily focuses on the issues of health, agriculture, and education, the increasing membership has brought the forestry issues into their purview. The new leadership, located in *upar* toli of the village, has, in turn, taken a populist approach vis-à-vis the forests. It, for example, does not demand forest guard duties of its members, and asks a nominal fee of Rupees 5 for each pole a member extracts from the forest. The final test on claims to leadership came when in 1993, the village was formally registered under the JFM arrangement and Lahru was made the secretary. However, with the lack of support from the forest department Lahru was left a leader only on papers. The recent input of lift irrigation by the NGO came as the last straw. The staunchest of the supporters in order to benefit from irrigation scheme have *migrated* and joined the loyalties of the NGO-run club membership.

VILLAGERS SUPPORTED LOCAL MANAGEMENT OF FOREST

The case of Ober demonstrates that villagers supported local management of forest. Lahru who could provide the leadership was able to challenge the traditional leader and was encouraged to expand his constitution to cover development issues. While, the onset of JFM, should have helped Lahru retain this leadership, villagers instead shifted their loyalty to the young leaders who were supported by a more responsive NGO. While, JFM makes it easier for the villagers to secure forest patches to meet their requirements, they clearly favour a situation where this security can be enlarged to other areas of rural life.

The Forestry Case of West Bengal

Bhagwati Chowk village, in Gopegarh Forest Beat in Midnapore East Forest Division, has a near uniform population make up, and is next to a relatively larger patch of forest. Poltu Singh has been the leader and spokesperson for the successful FPC since inception. Various features typical for Bengal mark the area. Many of the villagers live on land they got through left government's land re-distribution schemes. Poltu Singh has been a leader with *Panchayat* and much of his confidence comes from his familiarity with bureaucracy. However, currently the *panchayat* is under the schedule list and the council leader is a woman from schedule caste from the neighbouring village of Amratoli. Poltu Singh, however, remains a leader of his village FPC and his 'patch' of Gopegarh forest. The FPC, however, is less radical than Ober in Bihar. A 'successful' FPC in terms of good forest cover and with several instances of harvest and usufruct distribution, the committee is, in general, disinterested in forest activities on a day to day basis. It is active only during the official get together with Forest Department officials and infrequent committee meetings. The FPC has, however, created an 'exclusion zone' (similar to Ober) in the forest that lie under their jurisdiction, and takes up occasional fights with infiltrators from neighbouring village, often with villagers from same *panchayat*. It has often served as a platform to work in forestry micro-plans (when funds were available from the Forest Department) and with NGOs The impact that NGOs have had on rural development forestry is beyond the scope of this chapter. It is, nonetheless, worth mentioning that an NGO that could provide support in various areas of rural development were accepted more by the community than those who, for example, worked only to promote JFM activities.

There are, however, zones of anomaly in the region. The neighbouring villages of Amratoli and Phulpahari form another FPC and share a large block of forest. While Amratoli is in same *Panchayat* as Bhagwati Chowk, Phulpahari belongs to another *panchayat*. The population make-up is diverse and so are the forestry practices and dependency. The caste-Hindu settlements are better off than the rest of the population, and their women folk do not go to forest for collection of fuelwood. The political factions are apparent, and forestry is often compromised for political gains. The entire forest, for example, was cut down during the *panchayat* election in 1998.

Existing Social Capital

The norms of reciprocity and attitude towards trust are dynamic in nature, and choices are made in such a manner as to maximize the benefits to individuals. Once the common goals are within sight, the community builds on the existing social capital and makes adjustments to capitalize on the incentives. It seizes upon the new opportunities and does not mind risk breaking established conventions.

It also indicates the regulatory constraints of the system. A successful and popular village level forum cannot overspecialize in one issue such as forestry alone. A rigid dichotomization cannot be held, for example, between forestry and other rural development issues. Also, a populist agenda, as is common with new organizations such as NGO-club in Ober, may, in order to attract membership, harm the common pool resources like forests. Left to them, village institutions would, while seizing on the most attractive incentive, tend to compromise on old obligations, thus making the system inherently unstable. While in case of Bihar, on an NGO's involvement the villagers switch loyalty to capitalize on the opportunities for development, in Bengal *panchayat* leaders barter away forest timber in exchange for votes in an election. What is, however, unmistakable is that despite a marginal nature of FPCs (as compared to the established FD, *panchayats*, NGOs), these Forest Protection Committees often function to create a power-base and legitimacy of their own. The legitimacy that the FPC commands in Bhagwati Chowk is, for example, unambiguous. Poltu Singh despite his not being a council leader remains the leader in the eyes of Gopegarh villagers because of his formal control over a vital resource.

Informal Niche in JFM Arrangement

The section deals with informal In an effort to locate the informal practices, it is important to note that these activities are termed informal only when seen against the norms and rules that are laid out in the Joint Forest Management agreement. Many practices are not only essential to villagers' day to day's needs but have been in practice (noticed or disguised, contested or informally recognised) for a long period of time. Activities that find support in JFM's participatory atmosphere. Through time, the Indian State has exercised a near total control over its forest-land and forest citizens. The current participatory paradigm of sharing revenue and forest-usufruct through Joint Forest Management places this control (and polity) in an ambiguous position. The section argues that while the state still maintains a strong hold over both tenure and forest land practices, the new arrangement has made inroads for unprecedented concessions that allow rural people greater control over the village and protected (community) forests.

PROTECTION AND MANAGEMENT OF 'PROTECTED' FOREST

The FPCs, as a part of the JFM agreement, in return for protection and management of 'protected' forest would, apart from access to fuelwood, share a percentage of cash with the state on sale of timber. A range of concessions allowing collection of fuelwood and minor forest produce have existed (pre-dating the current participatory forestry) in both the States, though the relation between the Forest Department and citizens has been one of constant confrontation and suspicion. Bihar, traditionally, has enjoyed more rights over forest use than Bengal. There is no fee payable, in Bihar, for fuelwood collection or livestock-grazing, for example. The tribal population has also been entitled to a share of timber on occasional harvests. There is, however, no provision (including the JFM scenario) in either of the states to allow felling of trees other than those announced through government-circulars (sanctioned through official 'working plans') and carried out at the initiative of the Forest Department. The use of forest, nevertheless, is not limited to what is prescribed within the JFM agreement; villagers regularly extract logs of wood from protected forest for construction of houses and agricultural tools.

The forest bureaucracy is now no longer seen as a foe against whom village communities must unite against, often turning forest land into 'open access resources'. Instead, villagers are identifying more solidly with isolated forest patches that lie under the jurisdiction of village Forest Protection Committees. Forest lands are now demarcated and governed, for management purposes (protection, regeneration, harvesting, even silvicultural decisions), by these committees. The committees often decide on the exclusion of neighbouring village-communities from use of 'their' forests. To 'catch and punish the erring population' frequently forms the basis of protection of forests and to establish norms onto villagers from outside and even within the community itself.

JOINT FOREST MANAGEMENT IN INDIA AND BOUNDARY-MAKING

Colonial and post-colonial states have long attempted to extend and assert their control over forests. As part of their strategies, these states created specialised government institutions with policing authority over forests . These efforts involved the construction of new geographical boundaries through categorising different types of forest and designating different kinds of legal use. The Indian Forest Act of 1878, for example, defined three kinds of forest: reserved forests, protected forests and village forests. Such classificatory efforts largely disregarded the social and economic consequences of boundary-making processes for the inhabitants and users of the forests.

Since the 1990s, many developing countries have introduced alternative approaches that aim to involve local actors in forest management. The Joint Forest Management (JFM) approach in India is one such example. Since the inception of JFM in 1990, approximately 85,000 JFM forest protection committees have been put in operation, managing 27% of Indian forests (17.3 million ha of forestland) as of 2006. Given the previous history of top-down forest management in India, the dramatically increased interest of the Indian government in including local people in forest management is rather striking.

A variety of explanations have been offered for this new interest. Some have argued that the Indian JFM initiative illustrates the rising powers of civic society to shape equitable and democratic policies. Others have suggested that states have had few alternatives but 'to turn from coercion to consent' given the 'constant conflict between recalcitrant villagers and beleaguered forest staff in which both sides had been known to lose lives and limbs'. This argument aligns with the suggestion that the actual capacity of states to enforce regulations over forest resources dispersed over large territories may be quite limited, so that local involvement might in fact be the most effective way of achieving some measure of state control over such areas. Arun Agrawal (2005) has further argued that while JFM indeed functions as an extension of state power, this does not preclude the possibility that it might also create new opportunities for involved villages, and-at least potentially-facilitate the construction of new environmental subjectivities or social groups.

JFM shares some central assumptions with the influential framework of Common Pool Resource (CPR) theory. Most importantly JFM and CPR theory both emphasise the importance of establishing formal resource boundaries in order to achieve sustainable resource management. Of course boundaries create insides and outsides. Thus, new resource boundaries can have serious consequences for excluded villagers, if their livelihoods depend on resources available within these boundaries. However, this is not only a matter of which villages are included and excluded by JFM. Since villages are also internally heterogeneous in terms of gender, caste and livelihood activities , various social and economic consequences of JFM may arise even within villages located on the inside of resource use boundaries.

Drawing on a case study of three villages in Andhra Pradesh, two included and one excluded from JFM, this article considers the social implications of the natural resource use boundaries formalised as part of this policy initiative. The article shows how resource use boundaries interact with social categories such as caste, gender and livelihood occupation in ways that facilitate asymmetric distribution of costs and benefits among local people.

MAKING FOREST BOUNDARIES AND MAKING SOCIAL BOUNDARIES

As Scott made clear in *Seeing like a state,* mapping is central to the modern, administrative endeavour. He referred to the cadastral map as the 'crowning artifact' of the state ambition to 'measure, codify and simplify'. Scott noted that the mapping of forests was crucial to the territorialising ambition of colonial and post-colonial states, as it made the land legible and thereby more easily controllable.

The endeavour to map is also supported by present-day CPR theorists who promote boundary-making as a means to facilitate sustainable resource management. According to Agrawal (2003: 244), CPR theory can be characterised by its focus on the 'variations in the forms of property rights [that] make a difference in resource management outcomes'. With this focus, the CPR literature has analysed ways in which local collective action may hinder a 'tragedy of the commons' scenario in which nobody takes responsibility for shared common resources. Drawing on extensive empirical analyses, Ostrom (1990: 180) defines a series of design principles aimed to improve the endurance and efficacy of CPR management. Central among these principles is the insistence on the need to create clear biophysical boundaries for each local level unit involved in resource management.

Clear boundaries have also appealed to policy sensibilities. Formal demarcation of natural resource use boundaries has been adopted as a policy principle for JFM and other CBNRM initiatives, including community forestry in Nepal and Mexico , participatory forestry in Tanzania (Wily 2001), decentralised forest management in Bolivia , community based catchment management in northern Thailand, the CAMPFIRE (Communal Areas Management Programme for Indigenous Resources) project in eastern Zimbabwe, and community based resource management in Mozambique. These approaches devolve managerial responsibilities to local level units (which may vary from user groups, villages, or traditional councils to democratic local governments), and formalise their resource territory and rights to various resources.

Although CBNRM has generally been received with enthusiasm, critics have noted that the establishment of boundaries and the assignment of exclusive collective property rights to certain groups may have unintended and often adverse consequences. The ethnographical and geographical literature on the 'power of maps' argues that biophysical and territorial boundary-making often has social, economic and political effects for those whose livelihoods depend on crossing such boundaries to obtain needed resources. Focusing on 'cadastral politics' involved in local mapping exercises in Zimbabwe and Mozambique, Hughes (2001) highlights the interrelations between geographical and social boundaries. He remarks that harmonious relations and mutual benefits among involved actors such as governments and villages, which the notion of CBNRM implies, rarely exist. He criticises proponents of CBNRM for failing to take into account that bringing a resource area within the jurisdiction of one village is also a way of excluding neighbouring villages from access to the resource. A case study of forest land allocation in Vietnam's Central Highlands offers another example. Sikor and Thanh (2007) document that new forest management initiatives, even though they were considered beneficial and 'joint' from the point of view of one village, brought along negative consequences for another. These critical analyses call for detailed empirical studies of boundary-making through CBNRM initiatives, in order to understand the mixed social consequences that arise as natural resource use boundaries give rise to new social practices.

To illustrate the social transformations as a result of the newly created boundaries, Yeh introduces the phrase 'rearrangements of lived space'. Rearrangements of lived space are likely to occur both for those who are formally included as participants in CBNRM and their neighbours who are excluded. Rearrangements can be experienced as directly physical when villagers are prevented from accessing certain forest areas. But they can also be experienced through social and political effects when conflicts emerge as a consequence of the creation of new forest boundaries between villages that previously co-existed at relative ease. The processes set in motion by resource demarcations under CBNRM might therefore have ramifications aside from sustainable resource use at the level of the individual village.

Even though CPR theorists acknowledge the importance of conflict resolution mechanisms, the conflicted processes engendered by boundary-making have not been a focal point of analysis. In fact, CPR studies generally show little attentiveness to the social and political construction of natural resource use boundaries. This is likely because their focus is on defining principles required to achieve successful resource management rather than on tracing social implications in empirical detail. The consequence is that the boundaries of successful 'long-enduring CPR institutions' often come across as natural, rather than as having been made by specific people and institutions with particular agendas and interests. CPR analyses exhibit a certain analytical blindness to the consequences of*making* boundaries and especially to effects on those people falling *outside* boundaries. This is manifestly problematic when one is analysing *ongoing* processes such as JFM, where the new formalised boundaries are obviously neither natural nor neutral.

In an otherwise nuanced analysis of 'collective action, property rights, and decentralisation in resource use in India and Nepal', Agrawal and Ostrom reach the conclusion that 'at least at the local level, rural poor have benefited from the relaxation of state control over forest resources'. Yet, this conclusion is much too general. As our case shows, when the state transfers resource control into the hands of certain local groups, the rural poor are likely to be *differentially affected.*

To analyse how new formalised natural resource use boundaries rearrange lived space in and among villages, we take the notion of 'boundary-work' from sociologist Thomas Gieryn (1999). Gieryn proposes the notion of boundary-work as a tool with which to analyse the activities different actors carry out to *establish and maintain*-or *dissolve and redefine*-differences between social groups. Following Gieryn, we show how boundary-work can function in diverse ways: 1) as a *means of social control,* 2) as an *effort to protect the autonomy of specific groups,* and 3) as a *tool of strategic, practical action to procure specific benefits.*

We investigate the specific kinds of boundary-work through which villagers in the Medak district of Andhra Pradesh state engaged with, supported and resisted the JFM programme. Through this analysis, we aim to show how JFM engendered rearrangements of lived space in and between villages. The analysis aims to elucidate the intimate relations between the remaking of forest boundaries and the making and remaking of social boundaries in villages affected by JFM. By focusing on the boundary-work in relation to the creation of new resource boundaries, this study goes beyond others that concentrate on specifying success criteria for establishing resource boundaries while assuming that such boundaries, once well established, automatically have beneficial social or environmental effects. We argue that understanding boundary-work is significant in order to understand the success and challenges of the JFM policy, both in terms of ensuring sustainability of resources and in terms of promoting social equity and justice.

FOREST POLICIES BEFORE AND AFTER JFM

Before JFM, the aim of forest governance was to maximise economic profits from the forests under the jurisdiction of the Forest Department/state government. The Forest Department had the primary task of protecting the forest against intruders (ibid.). Under this arrangement, villagers had no *de jure* rights to make use of forests and the Forest Department officers exercised a policing power over forest territory.

The introduction of the National Forest Policy in 1988 brought about significant changes in forest governance. The new policy shifted the primary objective of governing forests from maximisation of revenue towards forest conservation. The policy also officially recognised the importance of incorporating the needs of local people in forest governance (ibid.). In 1990, the Ministry of the Environment and Forests issued a guideline for all states to adopt JFM. JFM facilitates the establishment of partnerships between the Forest Department and the forest protection committees that consist of local villagers. The committee is assigned with the task of managing and protecting the designated forest area against misuse (ibid.). In return, villagers who join the committee gain official rights to use some forest resources within the demarcated area for subsistence and income (ibid.).

FOREST BOUNDARIES BEFORE AND AFTER JFM

Prior to JFM, forest boundaries were drawn between state-owned forests and villagers who were restricted from accessing forests. In spite of the restrictions over villagers' use of forests, village informants explained that they had *de facto* access to forest resources. They used to enter forests to collect various forest resources by avoiding the timing and locations of patrols of forest guards employed by the Forest Department. This information corroborates arguments made by others, *e.g.*, Agrawal (2005) and Sundar (2001), that Indian forest managers have been relatively unsuccessful in preventing villagers from accessing protected forests, although they could avail themselves of hard measures *when* they caught offenders. According to village informants, Forest Department field officers were well aware of villagers' *de facto* use of forest resources. The officers sent forest guards as 'spies' to the villages to assess locations of newly constructed houses, details of goat shepherds (*e.g.*, number of herds, place of grazing, etc.), and details of collectors of tree poles and fuelwood. Based on such information, the officers demanded that each household pay bribes for their illegal activities and threatened villagers with fining offenders or taking them to court.

In these situations, both forest officers and villagers were involved in boundary-work. Thus, one central part of the job of forest officials was to maintain a strict boundary that kept villagers out of bounds of state-owned forests. This boundary-work took two of the forms identified by Gieryn (1999). First, it aimed to *protect the autonomy* of forests. Second, it functioned as a *source of social control* by restricting the movements and activities of villagers. Villagers, however, were able to circumvent the imposition of this boundary in diverse ways. Following Gieryn (1999) we see them as engaged in *strategic, practical action,* to procure various forest goods, breaching official boundaries and resisting both other villagers and villages in the process.

JFM led to a significant redefinition of these forest boundaries. As a part of JFM implementation, the Forest Department parcelled state-owned forests and assigned certain villages the task to co-manage forests. According to the account of the Additional Principal Chief Conservator of Forests in the Andhra Pradesh Forest Department, the field officers parcelled out forests and then used these parcels as the basis for assigning forests to villages. The Forest Department viewed the villages lying closest to the demarcated forests as most suited for enrolment and they approached these villages to establish forest protection committees. However, the Forest Department did not involve all villages. The Medak District Forest Officer explained that the reason for this was that not enough forest was available to share it among all the villages. Paradoxically-despite the general aim of JFM to support livelihoods-they made these new boundaries without consultation with villagers. The Forest Department also did not take into account customary use patterns of the demarcated forest areas.

SUPPORTING AND RESISTING THE JFM BOUNDARY

In 1997, the Mohammed Nagar village constituted a forest protection committee to govern the forest demarcated by the Forest Department. Two members (one male and one female) from each of the households belonging to the village constituted the general body. Those chosen selected 12 members representing each of the four castes to make up the management committee and management committee members nominated a chairperson from the upper caste. Under the leadership of the upper caste chairperson, Mohammed Nagar village held a general body meeting in the presence of all caste leaders to discuss how to protect their assigned forests while continuing to meet their own needs. First, they specified rules for kinds and quantities of forest resources that could be taken from the forests. It was decided that villagers were only allowed to collect dry fuelwood by head load and two tree poles per year free of charge. Beyond this, villagers were obliged to pay fees to the committee, *e.g.*, INR 30 for a cartload of fuelwood, and INR 300 for a cartload of tree poles. They also prohibited encroachment and the cutting of tree branches for fuelwood collection, and feeding goats. Second, they decided to employ two forest watchers from their own village. These watchers would patrol daily in the mornings and evenings in order to survey the forest and catch offenders. Further, the village offered 25% of the fines that were collected to those who caught the encroachers. The remaining 75% was to be deposited in the forest protection committee account. The rates of fines varied depending on the significance of forest offence, ranging from INR 100-300 for inhabitants of Mohammed Nagar village, and starting from INR 1,000 and up for other villages (in comparison, the average daily wage for forest work is INR 60-90).

Some of the rules and restrictions imposed by the village forest protection committee did not follow the rules defined by the Forest Department. According to official rules, the forest protection committee is not allowed to collect fines above INR 100 and all fines are to be handed over to the Forest Department. It is only after deposition that committees are entitled to 50% of the fines collected by the Forest Department. Realising that the rules had been changed, a Forest Department officer criticised the committee and demanded that the fines be paid to the Forest Department. The first chairman of the committee explained how the villagers responded to this demand by saying that they were risking their lives to protect their forest and arguing that "if you oppose our practice, why don't you patrol the forests and catch all the offenders yourself, then we will give you a fair share of fines". According to villagers, the officer gave up pursuing his claim and he left the forest protection committee to its own devices from then on. Similar instances of enforcing rules that were much tougher than those mandated by JFM policy were observed in neighbouring villages. This suggests that boundary-work relating to the protection of forest autonomy and enforcing social control was transferred from the Forest Department to the forest protection committee. Villagers prohibited from entering forest areas continued to engage in 'strategic, practical action' to access forest resources unnoticed by patrols on a daily basis. Yet, these 'subversive' activities were no longer directed against government officials but against their own and neighbouring villagers.

In spite of its commitment, the Mohammed Nagar forest protection committee faced numerous rule violations, both by people from the village and by outsiders. Forest offenders from the village were mostly from the tribes. They collected forest products without permission or illegally expanded their agricultural land to forest areas. Nevertheless, forest offences by outsiders constituted a far more serious problem. The forest protection committee aimed to strengthen the boundaries and caught many offenders from neighbouring villages. Gradually, several members of the middle caste became more active in forest protection activities, voluntarily taking jobs as forest guards. Their boundary-work consisted of patrolling forest boundaries and catching offenders. Although economically beneficial, this job was not without its hassles. As one middle caste informant explained "I have fought with so many violators from our own village as well as from neighbouring villages to stop the smuggling of fuelwood and tree poles from our forests, especially at the beginning.... It is a risky business since offenders often attempt to fight against us or even to murder us".

Yet, after several years, the persistent efforts to enforce the boundaries had resulted in a reduced number of forest offences. The number of recorded offences in the Mohammed Nagar village dropped from 20 in 1998 to three in 2007. While this official record may show only a fraction of actual number of offences, it indicates a clear reduction of major offences in Mohammed Nagar. As a result of the reduced number of offences and restriction over forest use, Mohammed Nagar's forest started to regenerate. Forest regeneration also meant that villagers started to benefit from their exclusive and increased access to forest resources. Their benefits from JFM included permission fees from its own villagers and fines from villagers inside and outside the village. During the period between 1997 and 2002, for instance, the forest protection committee collected a total of INR 1,64,861.

SOME WELL-ENTRENCHED BOUNDARIES: CASTE, GENDER AND LIVELIHOOD

The case of Mohammed Nagar village was a success story for JFM in that forest conditions improved and villagers gained almost exclusive access to their forests. We take a closer look at the asymmetric distribution of costs and benefits arising from JFM within Mohammed Nagar village. The asymmetries we identify are predominantly related to well-entrenched social categories defined by caste, gender, and livelihood. Whereas the previous sections highlighted how new forest boundaries may create new types of boundary-work among villages, this section documents the difficulties of changing inflexible social categories and boundaries relating to caste, gender and main livelihood activities by means of new policies. This applies even to policies such as JFM that specifically aim to address questions such as gender inclusion and equality. How does boundary-work relating to JFM interact with established social categories, and with what consequences?

The forest protection committee devised a set of rules that would apply equally to all villagers. Yet, these rules had the largest impact on those whose livelihoods were highly dependent on forest products: the lower caste groups such as tribal people and the landless. According to Seva Sangam, a local NGO which conducted a survey of household collection and sale of fuelwood of six villages including Mohammed Nagar in 2007, one thirds of the upper and middle caste people and around 10% of the scheduled caste purchase fuelwood from tribal people rather than collecting it themselves. The rest of the castes collect fuelwood only for domestic use (about six to nine kg of fuelwood per day). In contrast, tribal groups collect more fuelwood than other castes for domestic use (about eight to fifteen kg of fuelwood per day). 10-15% of tribal groups collect additional fuelwood for sale to other castes. Some tribal people are in the wood cutting business which requires a large amount of tree poles. Since the new rules were introduced, they have had to pay additional fees to the committee to collect beyond the permitted amounts. In addition, tribal people who live adjacent to forests used to increase their livelihood opportunities by gradually cutting forest to expand their agricultural lands. Prior to JFM, this boundary-changing practice was prohibited by the Forest Department. After JFM was initiated, however, local villagers themselves enforced a ban on the practice.

For these reasons the JFM restrictions on permitted amounts of collection of fuelwood, tree poles and on encroachment created more pressure on tribes than on other castes. For many tribal people, the benefits of the new forest protection system were far from obvious. 21% of 74 tribal informants perceived JFM to negatively affect their ability to derive benefits from the forest. One tribal villager also explained that "fighting with the government is easy but fighting with our own villagers is very difficult". According to him, the reason for this is that the villagers (represented by the committee) exercise greater authority in checking his activities than the Forest Department. His remark suggests that boundary-work relating to forest access and use may intensify with JFM, since in a Foucaultian optic [similar to the one adopted by Agrawal (2005)] monitoring and surveillance can be much more continuous and thorough when conducted by villagers than when carried out by Forest Department officers who employ only sporadic checks. Another consequence of the implementation of collective rules was the emergence of new social categories. Gradually villagers had begun to refer to their own villagers who broke rules and neighbouring villagers that took forest resources from their area as 'forest offenders or intruders'. Likewise terms such as 'rule-followers', 'enforcers', and 'violators of rules' came into usage. In the case of their co-villagers, these new categories largely corresponded with caste. During interviews, village informants from the upper, middle and scheduled castes often referred to themselves as the main followers and enforcers of the rules, and referred to tribal people as the main violators.

Just as punishments for rule violation were asymmetrically distributed, so were benefits from the common revenue generated by JFM. The forest protection committee generated significant common revenue from fines and permits. The committee decided to use the revenue for the construction of three Hindu temples in the main village. Even though the committee, in principle, represents all castes, these temples benefited higher caste people only (since the 'untouchable' scheduled caste people were not permitted to enter the temples and tribal people are not Hindu). This decision which clearly favoured the higher castes was passed presumably because the committee chairman was from the upper caste and a majority of management committee members belonged to higher castes (eight out of 12 management committee members).

Aside from caste differences, no social categorisation in the villages is as important as gender. Women are the main stakeholders in forest management as they are collectors of fuelwood and NTFPs. At the same time, women are socially disadvantaged as they have little opportunity or inclination to express their views in public. To ensure equal participation, the Andhra Pradesh government issued a government order that mandated more than half the general body and the management committee was to be constituted by women (Government of Andhra Pradesh 2002). Nevertheless, actual participation of women was nominal, as their husbands generally attended meetings on their behalf. 61% of 109 female respondents answered that they *never* attended general meetings (only 28% of 113 male respondents gave the same answer). The main reasons for non-attendance mentioned by women were that 'they are busy with household work', that 'they feel uncomfortable attending male-dominated meetings', and that 'it is not accepted by men that women attend meetings or speak up in a male-dominated crowd'. A scheduled caste widow from the second term management committee complained: "I was always the only female who attended management committee meetings. How can I make my opinions heard when all the other women are absent?" As a result, the women's point of view was not reflected in decisions and rules on forest related activities. Correlatively, there is a notable difference between male and female respondents' perceptions over effects of JFM on benefits from forests. Whereas 82% of 111 male informants perceived positive effects, only 38% of 107 female informants perceived such effects.

Finally, asymmetric distribution of costs and benefits within JFM villages also relate to main livelihoods. To exemplify this point, we draw on the case of Kanchanpally village where the new forest boundary adversely impacted many goat herders. Kanchanpally village joined JFM in 1998 and introduced tough restrictions on forest use. In the village, around 50 out of 382 households depended on goat herding as their principal livelihood activity. Goat herders had followed the customary practice to move goats approximately 20-30 km on a daily basis to allow them to graze. For this reason the Forest Department considered goats harmful for the regeneration of forests and aimed to regulate the size of herds, and limit grazing by collecting fees per goat and fining herders. Since JFM, however, forest protection committees have overtaken this role.

The committees prohibited goat herders from their own villages from cutting tree branches and herders from outside the village from grazing their goats. Due to these restrictions, herders in Kanchanpally experienced increasing difficulties since the forests were insufficient to meet the dietary requirement of the animals. This meant that herders were forced to either ignore the restrictions on cutting branches or move to neighbouring forests governed by other forest protection committees. 12 of the 26 goat herders responded that they had been caught either by their own or by neighbouring forest protection committees. Altogether they had paid a sum of INR 80,550 in fines during the period 1998 to 2008. Due to strict regulations and fine charges, the Kanchanpally committee reported that the population of goats declined from around 2,500 to 1,000 between 1998 and 2008. Unsurprisingly a majority (81%) of these 26 respondents perceived negative effects of JFM on the benefits they derived from forests. Conversely, the forest protection committee viewed goat herders as both the main violators and the main source of revenue.

JFM has both benefited and harmed villagers who were inside the formalisation of natural resource use boundaries. Even as JFM formally includes lower caste people and women, social categorisations continue to impose social, economic and political boundaries that work against equality and inclusion. Thus, actual consequences of JFM result from *the interactions between natural resource use boundaries and social boundaries defined by caste, gender and livelihood*. These are social categories that show little propensity to change, except perhaps very gradually, and which, therefore, are*more likely to modify new policy initiatives than vice versa*.

VILLAGE BOUNDARIES AND THE REARRANGEMENTS OF LIVED SPACE

We have indicated some of the varied ways in which JFM become enmeshed with social boundaries, that are difficult to change, but which impose an asymmetric distribution of costs and benefits, even *among* JFM villagers. However, the consequence of being left on the outside of new forest boundaries is far more damaging. As a result of JFM, villages that were not officially recognised as managers of forest lands quickly began to face difficulties. Their previous access to forests was blocked by village guards hired by neighbouring forest protection committees who demanded payment of large fines. Yet, the people of such villages had little choice than to continue to transgress these boundaries, involuntarily adapting to the emerging social category of 'forest offender'.

Antharam village exemplifies this situation. As JFM commenced, a Forest Department officer tried to persuade the village to join JFM. The political head explained that the village decided to refrain from joining because villagers had no idea of the consequences. As a result the officer allocated all nearby forests to neighbouring villages including Mohammed Nagar, Salbatoor, and Kanwaram. As forest protection activities began to be implemented by JFM villages, this caused difficulties for Antharam. In order to avoid being caught and fined by forest protection committees, they attempted to undermine the boundary, by sneaking into adjacent forests during the night in order to gather fuelwood. Antharam villagers perceived the JFM as unfairly restricting their ability to procure necessary forest products and view neighbouring village guards as purely economically motivated. One expression of a sentiment often encountered was that "our neighbours are not after protection of forests but after fines they collect from us". 22 out of 55 surveyed households reported that they had been caught by the forest protection committee of other villages between 1998 and 2008. Together they had paid fines of INR 48,600. All (100%) of the surveyed 55 households in this village viewed the impacts of JFM as 'negative' regardless of their caste and gender. Antharam is, of course, not the only village excluded from JFM.

Antharam village did not remain passive in the face of these problems. The village approached Forest Department officers numerous times to request allocation of a parcel of forest land for their management and use. Village representatives also contacted Salbatoor village about the possibility of becoming included in their forest protection committee. These proposals were rejected. In conversations with Forest Department officers about the difficulties of Antharam village, we were repeatedly told that nothing could be done since 'all forests had been allocated'.

This example indicates that newly made natural resource use boundaries may grow rigid very quickly. Although formalisation of boundaries may be a key success factor in sustainable resource management, the situation of Antharam village indicates that problematic socio-economic consequences may follow from the rapid naturalisation of such boundaries. At this point we can return to Agrawal and Ostrom's (2001) estimation that the rural poor have generally benefited from JFM. This may certainly be the case for (some) villagers *included* in the initiative. However, *de facto* not everyone is included or can be included in JFM. This raises important questions about the equity and distributive justice of JFM.

PARTICIPATORY POLICIES IN THE REALM OF NATURAL RESOURCE MANAGEMENT

A series of new incentives and power-relations have been created because of participatory policies in the realm of natural resource management. Encroachment of forest land, for example, for agriculture and less so for homestead purposes is not uncommon. Understandably, while many of the encroachments have existed before JFM got implaced, with the current management practice, the Forest Department finds it difficult to wrest the land out of these villagers. The encroachment is mainly by people who have leverage within the village community, and are often active members of committee. Finally, in more extreme cases, defaulters have used the insularity derived from powers and legitimacy of the new committee, to sell timber for commercial purposes. The participatory schemes have somewhat also resulted in limiting both the spheres of influence and activities of the forest department. The policing, for example, is now restricted to 'non-negotiated' territories, areas that are not put under a village's jurisdiction: roads, forest-corridors, reserved-category forests and *haat*, the local-markets (where forest-usufruct is bartered). While in Bihar, the petty officers still patrol forest in uniforms, they enter villages only on being reported or in instances of outright felling of timber. The villagers enjoy a situation where they can, by showing accountability to community-forums, get away with practices that are not allowed within the JFM framework. Another factor which is common in both the states is the initiation of protection of forest by communities well before the state sponsored participatory policies took shape (often around 1985, encouraged by Social Forestry). The villagers vividly remember the hardships they have faced on many occasions due to unrestrained felling (often by the State through planned coupe-felling), and there is a constant reminder, that unless some manner of restraint is kept, forests will then turn yet again into Open Access Resource. While the State do not provide the FPCs with any regulatory powers, many of the norms have been imposed through practice such as extracting fines from the defaulters.

INDIGENOUS FORESTRY: PERCEPTIONS & METHODS

The working methods employed in informal activities and transactions are not hard to ascertain. Since the onset of participatory schemes, there has been decrease in policing and violent confrontations between villagers and forest officials. This marked *withdrawal* of forest department staff from actual resource area has prompted many inherent, albeit veiled, activities to come to the fore. JFM-led village-forums often 'regulate' these practices, without adhering to the boundaries (marking out concessions & rights) that the management agreement has prescribed.

Creation of 'Zones of Exclusion'

A key way the FPCs have laboured to protect local forest, is through creation of 'zones of exclusion'. The forest that lie under a village's jurisdiction are marked 'out of bounds' for people outside their community, and, if violated, fights are taken up with infiltrators from neighbouring villages. Again, the boundary that constitute these zones is not always that have been delineated by the FD. In Bihar, for example, the villagers, depending on their perception, mark out territories in forest-land, as their *rakhat*, or an area that is under protection, to exclude outsiders. Within the community setting, however, runs a noteworthy egalitarianism: not only they maintain a understanding of a universal need of extracting fuelwood (usually of fallen twigs, branches, of green shrubs, and at times of live trees, branches) but there also exists a consent, in principle at least, for extraction of timber for meeting essential needs like construction and repair of houses and for agriculture. These activities generally need a 'clearance' within the forum or carry understanding of the local leader. In Bhagwati Chowk, for example, it suffices to have the understanding of Poltu Singh (though many others would be informed). In Ober, the NGO-run club authorizes extraction of trees for a small donation.

The villagers own reading into Joint Forest Management systems is, at best, vague, and their activities, in the assumed participatory atmosphere, are instead informed through their own interpretations. They, though, are aware of the committees' responsibilities to protect forests (chiefly, from 'others') through guard-duties (and self-restrain), the nature of rights they are to have in return is uncertain. It is commonly understood that forest is not open to indiscriminate logging, but would, however, provide for the fuel and timber to cover basic needs. The practices are grounded on strong notions that forest belongs to the entire village community, and while rights on its land-tenures rest with the government, the villagers can not be alienated from their day to day use, 'the *sircar*, after all, can not take forests away', the villages reason. In their interaction with the Forest Department, the villagers frequently report of disappointment. They complain of not being rewarded in instances when they reported defaulters to the department, nor have they received material-support to aid them guard forest. Importantly, Forest Department seen as a part of the 'government', is expected to bring in development projects, while this is seldom the case. Another requirement of the JFM agreement, that villagers would replace the Forest Department with provision of guarding duties, flagged after initial bouts of enthusiasm.

Finally, the principal way JFM departed from previous participatory initiatives, by promising cash returns to FPC members, has produced mixed results. In Bhagwati Chowk with 37 households and nearly 150 hectares of forest under the FPC's jurisdiction, the two annual harvests that were carried out by the FD have fetched each of the households with significant amount of cash. This has rarely been the case in other situations. Amratoli, the neigbouring FPC with 240 hectares of forest and 246 households has a much lower forest/household ratio and cash return on harvest is hardly an incentive for them to institutionalize JFM practices. Bihar, on the other hand has yet to carry out its first harvest under JFM agreement. Here, and in instances where the cash-incentive is not sufficient, the reason for involvement lies more in the actual control of the resource and also in access, this newly found base provides, to stronger institutions.

Gender Divide in Community Forestry

Division of labour in extraction of forest usufruct is sharp. Majority of fuelwood extraction is carried out, often on a daily basis, by women members of the household. Forest in some ways is their exclusive work-zone, so much so that men find it socially awkward to scrounge forest-floors. The role interchanges when men have to cut trees for timber, for construction of houses and agricultural implements. Despite woman foraying into forest more frequently, the control that men exercise over forestry practices is unmistakable. Men think they can always stop women if they go *haywire*. A usual explanation by women, for extraction of wood, is, '*pooch kar late hain*, we ask before we bring forest usufruct'. The case changes slightly in Bengal, where effeminated men do, at times, collect fuelwood and even barter them in markets in the nearby township. Despite a less profound role in forest usufruct collection, men (as, also, shown in her thesis by Sarah Jewitt, and quite in contradiction to Vandana Shiva's eco-feminist principles) demonstrate considerable knowledge of time, species and pathways that regulate choices, in forest, on part of their womenfolk. While forest provides women with a *zenana* womb, where they socialize with other female members of community, it remains an 'alien' quarter where familiarity is seldom past a point, never beyond a time: what with herds of elephants, and one respondent even saying 'a small tiger' that are present in forests. Women rarely venture out on their own.

7

Climate Change Impacts on Forest Ecosystems

Reviews by Lucier *et al.*, (2009) and Fishlin *et al.*, (2009) on detected impacts, vulnerability and projected impacts of climate change on forests found that impacts varied across the continents with some forest types being more vulnerable than others. Impacts included increased growth, increased frequency and intensity of fires, pests and diseases and a potential increase in the severity of extreme weather events (*e.g.* droughts, rainstorms and wind). Human activities, including forest conservation, protection and management practices, interact with climate change and often make it difficult to distinguish between the causes of changes observed and projected. Deforestation and fires in the Amazon region, for example, form a vicious circle with climate change, with the potential to degrade up to 55% of the Amazon rain forests.

In this book, observed and projected changes in climate and weather conditions and their impacts on forest composition, structure, diversity and processes for the major forest types in different parts of the world are discussed.

FOREST CONDITIONS AND AREA

The area covered by forests is very likely to change under climate change, with shifts occurring between forest types due to changing temperature and precipitation regimes, while in some regions, forest area is expected to expand (*e.g.* temperate regions) and in others to contract (*e.g.* boreal, tropical and mountain forests). Such changes have been occurring in the past following the natural changes in temperature and precipitation that accompanied the different ice ages. Currently, however, it is very difficult to separate forest area change due to climate change from area changes due to other factors.

Globally, planted forests and natural regeneration have increased the forest areas in the United States, Europe, China, and some countries in Latin America and the Caribbean *e.g.* Chile, Uruguay, Cuba and Costa Rica. On the other hand, some countries in Africa, Asia and the Pacific and the tropical countries of Latin America continue to be subject to deforestation, mainly due to conversion to small- and large-scale agriculture and livestock while deforestation in the boreal forests of Siberia is mainly due to forest fires. Although the boreal forests are expected to move northward, temperate forests are expected to increase their area northward to a greater extent than the boreal forests, thus reducing the total area of boreal forests.

In the future, it is expected that the combination of climate change, land use conversion and un-sustainable land use practices will interact. Changes in water availability are considered to be a key factor for the survival and growth of many forest species, although the response to prolonged droughts will vary among species and also among different varieties of the same species. Climate change will increase the risk of frequent and more intense fires, especially where changing climate is accompanied by lower precipitation or longer dry periods as in the boreal , Mediterranean and sub-tropical forests and traditional land clearing practices as in the Amazon. In the northern Atlantic region of Nicaragua, for example, Rodriquez *et al.*, (2001) found that the combination of the amount of rainfall during the previous three months and the average monthly temperature of the current month showed a strong relation with 64% of the fires between 1996 and 1999.

Although data are not conclusive, it is expected that frequency of strong hurricanes will increase in hurricane prone areas such as Central America and the Asia Pacific region. Hurricanes may destroy forest areas completely or cause heavy degradation. If left untouched, however, such areas will ecologically recover over time (*e.g.* Vandermeer *et al.*, 2000; Vandermeer *et al.*, 2001), albeit slow in terms of biomass. The main effect is likely to be economic (infrastructure, crops and timber lost) and social (lost lives and livelihoods). Together with land use changes, however, the effects may be much longer lasting and devastating - degraded and young forests are easily converted into agricultural land and pastures.

HEALTH AND VITALITY OF CLIMATE CHANGE

Climate change may have profound impacts on the health and vitality of the world's forests. In some cases, vitality may increase due to a combination of a more favourable climate for growth and CO2 fertilization. In most cases however, increasing temperatures favour the growth of insect populations that is detrimental to the health of forests.

This is more likely to occur in forests dominated by few tree species or where specific temperatures or moisture levels control insect populations. For example, the spread of the mountain pine beetle, Dendroctonus ponderosae, in boreal forests, has been largely attributed to the absence of consistently low temperatures over a long period of time, which allowed an existing outbreak to spread across montane areas and into the colder boreal forests. Similarly, Finland is expecting an increase in infestation of root and bud rots in their coniferous forests, due to the spread of a virulent fungus, Heterobasidion parviporum, favoured by longer harvesting periods, increased storm damage and longer spore production season. In the tropics, on the other hand, increased warming reduces the life cycle of many insect pests, while at the same time increased fire damage makes trees more susceptible to insect attacks and vice versa.

Biological diversity of Climate

Species growth and survival depends for a large part on climate variables. Most species have a particular climatic range within which they grow best, are competitive and are able to adapt to slight environmental changes and respond to insect attacks, diseases and other adverse environmental and human influences. Many of the ecological processes that are needed for tree and other plant and animal species to live together are influenced by climatic conditions. The importance of climate for forest ecosystems and their composition and diversity is exemplified by the various global and regional vegetation classifications. The Holdridge ecological life zones , are limited by temperature, precipitation and humidity. Several researchers have attempted to estimate the impact of climate change on the forests of Central America, based on estimated shifts of the life zone boundaries. Such studies, however, fall short of projecting real changes that may occur, since geographical shifts due to climate change are likely to occur on an individual species level, rather than on forest type level. This is mainly because some species will be able to adapt better to changing conditions than others, resulting in changes of composition of forest types, rather than geographic shifts of forest types.

In general, many species have a tendency to move to higher latitudes or higher altitudes. Lucier *et al.*, (2009) in their revision of climate change impacts on forests, found reports of phenological changes in a number of species, with more and greater changes observed in higher latitudes. Common changes observed were changing flowering times and changing time of bud break, affecting productivity and carbon sequestration potential. Phenological changes observed in oak , apple and pears and a range of 29 Mediterranean species , did not affect ecosystem processes other than bringing them a few days forward, although such behaviour was easier to predict in insect-pollinated species than in wind-pollinated species. Ecological processes such as pollination, flowering and fruit setting may be more affected in tropical systems, by changes in the phenological cycles because species interactions may be more complex and involve more than one species, while at the same time seasonality is not as clearly marked.

FOREST ECOSYSTEM SERVICES AND UNDERLYING PROCESSES

Following the Millennium Ecosystem Assessment report (Millennium Ecosystem Assessment, 2005), and forest ecosystem services are defined as the benefits that people obtain from ecosystems. While many ecosystem services can be identified and are often grouped into four broad types of services , only those services with well documented evidence of their management and their relation with climate change and human well-being.

PRODUCTIVITY OF CLIMATE CHANGE

The impact of climate change on productivity varies according to geographic area, species, stand composition, tree age, soils (in particular water holding capacity), effects of CO_2 and nitrogen fertilization and interactions between any of these factors. Some of the changes may be temporal, reverting once saturation levels have been reached. This is projected to be the case for water availability, where reduction of water generally reduces plant growth but in areas of water surplus may initially increase growth when waterlogging is being reduced. Similar reactions have been noted for CO_2 and nitrogen fertilization as well as temperature increases. In general, productivity was found to increase with rising temperatures in most forest areas, including the Amazon, probably due to CO_2 fertilization. However, in contrast to temperate areas, production increases in tropical forests will be temporal and will decrease once CO_2 saturation levels have been reached. Some studies have already registered decreasing growth rates in tropical forests. Water deficits over extended periods have also been shown to decrease productivity and may be the cause for the declined productivity recorded by the studies above. Some authors argue that based on paleontological evidence this may not result in the forest dieback often mentioned in connection to expected changes in the Amazon region. Natural disturbances often decrease forest area, but through the damage they cause to standing trees, they may also decrease productivity.

CARBON STORAGE AND SEQUESTRATION IN FOREST

There is an important interaction between carbon storage and sequestration by forests and changing temperatures and precipitation. On the one hand, the more carbon is stored in forests; less will be in the atmosphere. Increasing this stock will thus contribute to reducing the rate at which the global temperature is increasing. This relation has become extremely important in the climate change discussions and many tropical countries are preparing themselves to reduce emissions and increase forest carbon stock in order to capture part of the funding pledged for GHG emissions reductions. In Costa Rica, recognition of this service led to the implementation of innovative financing mechanisms for forest management, planted forests and conservation during the mid-nineties. This has led to increased efforts to ascertain the extent and content of the existing natural and planted forests.

On the other hand, increasing temperatures, longer dry seasons and increasing CO2 concentrations in the atmosphere in the long term, are expected to reduce the capacity of forests to store and sequester carbon, possibly converting forests from carbon sinks to carbon sources. Since carbon sequestration depends on productivity, all factors that affect productivity will also affect carbon sequestration. In addition, in the short term, increasing temperatures may reduce carbon storage capacity, although the effect may vary depending on the season in temperate regions. Early spring warming, for example, has been found to increase carbon sequestration of terrestrial ecosystems, while early autumn warming increased respiration more than sequestration.

Soil and Water Protection

Forests have long been recognized as contributing to water and soil protection and in several countries this has been translated into systems that pay for these services. Their positive influence on water regulation, however, is still discussed by foresters and hydrologists. The role of water regulation and soil protection may become increasingly important under climate change conditions. However, the capacity of forests to fulfil this role may be affected by the changing conditions.

Reductions in rainy season flows and increases in dry season flows are of little value when total annual rainfall is low and significantly evaporated and absorbed by forests. In areas with frequent fog, the absorption of water by trees from the clouds (horizontal rain) may contribute significantly to the total amount of rainfall. The palaeoecological study of Amazon vegetation changes , indicated that in cloud forest areas, where trees often are submerged in fog, warming may cause the clouds to rise above the trees. This will reduce the potential for horizontal precipitation.

Multiple Socioeconomic Benefits

In some areas, climate change may increase growth, while in others decreases are expected. While the expected global increase in wood production may lower prices, benefitting consumers, the combination of lower prices and regionally differentiated effects on productivity will cause differentiated effects on timber harvest related income and employment. The same authors project rises in timber production of up to 50% in all continents, except for Australia and New Zealand. However, most of this increase is expected to come from plantations, with increasingly shorter rotations and is therefore likely to be distributed unevenly amongst the continents. In South America, where greatest increase is expected, current plantation production is concentrated in southern Brazil, Argentina, Uruguay and Chile. Natural forests are found in the tropical regions of the continent, where forest dieback may decrease timber production.

Harvests of non wood forest products (NWFP) have three major functions: provision of part of the daily necessities of forest dependent people, off-farm income and a safety net in times of adverse conditions for agricultural production. Osman-Elasha *et al.*, (2009) suggest that climate change will have impacts on the productivity of NWFPs and that NWFP users will largely be impacted through increased pressure on forest products from people that look for emergency supplies or alternative ways of income. The latter is likely to occur in areas of high poverty, high dependence on NWFPS and increased frequency and intensity of extreme climate events and other natural disturbances, such as pests, diseases and fires. The impacts of climate change on the provision of these products and the subsequent socioeconomic effects, however, require more studies.

Climate change impacts on cultural and recreational services of forests have also been little studied and are difficult to measure, in particular for those services that by themselves are difficult to measure. Osman-Elasha *et al.*, (2009) report some studies on well defined recreational services, such as skiing in mountainous areas, where skiing at lower altitudes is likely to be affected by temperature increases. Recreational values placed on forests are usually local and unfortunately in most countries no reliable climate change projections have been made at such a scale. The same authors indicate that the effect of climate change on forest biodiversity and structure in Africa and the subsequent effect on attractiveness for tourists of many of the national parks need to be further studied.

CLIMATE CHANGE TO FOREST MANAGEMENT AND ITS TECHNOLOGY

Climate change poses new challenges, opportunities and constraints for forest management.

These include changes in:

- the natural environment, which is the basis for forest management;
- the socioeconomic environment, particularly where local people depend heavily on the goods and services from forest ecosystems;
- international and national policies and legislation, such as REDD+ agreements, land tenure agreements;
- the markets, such as the carbon market, and;
- relations between different stakeholder groups, exemplified by the increased recognition of the tenure and intellectual rights of Indigenous Peoples.

These changes pose challenges for forest users. In some cases, they may be opportunities while in other cases they may constraints. This will depend on the user, type of use, geographic location and the current local socioeconomic and political situation. The possible implications of these changes for the management of forests for different objectives will be discussed in the following subsections, following the seven thematic elements for SFM endorsed by FAO.

CHANGES IN THE NATURAL ENVIRONMENT

STRENGTHEN ADAPTIVE CAPACITY OF FORESTS

Negatively affect forests and many of their plant and animal species. In addition, they may negatively affect the availability of other resources, necessary for species survival.

Current forest composition and structure are however, the result of past changes in climate and shows that forests and their species have an inherent capacity to adapt to change. The main differences of current climate change with historic changes are the increased rate of these changes and the degraded and fragmented state of the remaining forests, which reduces the capacity of the species and ecosystems to adapt. The challenge is to help species and ecosystems to adapt to climate change while at the same time ensuring that ecosystem services are maintained. This will require the identification of the changes to which the forest will need to adapt.

Locally, changes may be disastrous, unless climate, ecosystem and species changes are accompanied by adjustments in the local social and economic systems. For example, increased occurrence of severe fires will require greater collective action to prevent fires as well as improved weather and fire danger forecast services . Companies producing furniture of high value species from natural forests, whose natural regeneration under changed climate conditions has become increasingly difficult, may have to change geographic range for their inputs, or change to other species and/or other processing procedures. Communities and private landowners depending on local forests may have to change livelihoods after severe hurricane damage.

Nationally or at the landscape level, changes may be slower and less disastrous in the short term. New challenges include the identification of those species groups and ecological processes that are essential for the most important ecosystem services. This would include in most cases identification of water catchment areas (hydrogeology) and the role of forests in maintaining water quality and quantity. It will be important to increase the probability that changing ecosystems will continue to provide the important services and goods. In particular, ecosystems in geographic locations at the extreme limits of climatically well-defined areas, such as mountainous forests, rangelands and boreal forests, are likely to be severely affected and may disappear. Some authors suggest that maintaining functional diversity and 6 composition will preserve ecosystem services, while others found that different functional groups will react differently to environmental changes , indicating that climate change may favour some functional groups over others. More research is needed however, to identify those functional groups essential for the desired ecosystem services and goods in particular areas and to understand how these can be conserved and protected.

Reduce Risk and Intensity of Pest, Disease and Fire Outbreaks

Reducing the climate induced risk of pests, diseases and fire outbreaks, in particular, in dry areas and less diverse forests will be a major environmental challenge. Breeding of more resistant or more resilient varieties is a medium to long-term solution for plantation species, although, that introduces new risks because strengthening the adaptive capacity of a species for one trait may weaken it to other traits. Identifying species for their "realized fitness" - for example, varieties of a species that survived insect attacks, diseases or fires, similar to the expected events in a particular region - and then facilitating their migration to the area of interest, may be another strategy. In both cases, identification of the traits that will increase resistance or resilience will be important as will be replicating those traits over generations and successfully introducing the species or varieties in the area of interest, without introducing new problems (such as undesired invasion).

Predicting future changes in pest and disease outbreaks and adjusting management accordingly is another option, which requires the development and validation of models that reliably predict impacts under different climate and management scenarios. A further option is the identification and implementation of forest management systems that are known or thought to reduce the risks of pests, diseases and/or fires.

While there are several well known means to protect forests and plantations, in many cases these are not applied for a variety of reasons , or are not applied to those forests most in need. The challenges are to identify and address the reasons for the lack of application of management techniques and to adjust management options to the threats in a participatory, socially and economically acceptable manner.

CHANGES IN SOCIOECONOMIC ENVIRONMENT

Risk of Migration into Forest Areas

Climate change will affect all people but in particular, rural people that depend on nature for their livelihoods, and poverty stricken communities in the urban-rural interface that are often subjected to the consequences of extreme weather events. Climate change is expected to change the aptitude of lands for specific crops, cause problems of droughts, fire and flooding and may drive many people from their lands. These people are likely to either go to cities to look for jobs, often adding to urban poverty, or to other rural areas to look for other lands where they may be able to continue their agricultural livelihoods or find employment in the agricultural sector. The surge of interest in fuels from biomass (*e.g.* corn, sugarcane and oil palm) adds another dimension to this migration. The purchasing of land, often based on speculation, in the hope of selling later for higher prices to investors interested in biofuel production, may cause migration. The expected high incomes from biofuels may also motivate landowners to convert their forests into energy plantations , oftentimes in an unsustainable manner. On the other hand, if well planned, biofuels could also help avoid or reduce migration by providing off-farm employment.

Forest use values, even in the case of the most successful enterprises, will not be able to compete with oil palm or other energy crops in those lands suitable for the crops. Legal definition of user and owner rights of forest areas and the mechanisms to defend those rights will be important elements of strategies to prevent unauthorized entrance into forests. Market mechanisms that restrict trade of products from companies that do not show social and environmental responsibility in their production and purchase policies may be another strategy. An individual forest user or owner will find it difficult to influence legislation, their implementation or the way that markets function. Collaboration with other stakeholders, neighbours, value chain members, and state administrators will be essential to the development of adequate measures to reduce the conversion and degradation of forests. Forest users and owners, however, have a longstanding tradition of independence and in the past have not shown tendencies to such collaboration. Lack of trust (often justified), has often hampered relations between different stakeholders in the forest and environmental sectors. Building sufficient trust to facilitate collaboration may be the biggest challenge of all for future forest management and needs the collaboration of all actors involved.

FOREST ECOSYSTEM SERVICES BY LOCAL PEOPLE

Climate change is expected to increase the frequency and intensity of extreme weather events, such as hurricanes, torrential rains and droughts. Rural people often depend on emergency supplies during or just after such events. Forests, in many cases in the past, have provided such emergency supplies or safety nets *e.g.* wood for construction and repair of houses, woodfuel for cooking and fruits and other food to replace the lost crops. The need for these safety nets will further increase when climate change increases the loss of crops. Indigenous groups are often vulnerable to extreme events, especially those events that restrict access to the outside world and markets. However, in such cases, they can usually find sufficient emergency supplies from within the forest until access is restored. In addition, more people have become aware of the different ecosystem services and want to use such services even under non-extreme weather conditions.

Forests as regulators of water quality and quantity have become ever more important, in particular, in areas with frequent droughts and/or frequent torrential rains that may cause erosion, sedimentation and flooding. In Central America, this function may be one of the main reasons for forest protection or restoration by private landowners even though it is possibly based on an erroneous perception of the benefits of the forest, since such functions may not be beneficial in some climate and soil conditions. The impact of climate change on this ecosystem service, however, is still not very well understood, since different species, different environmental and geological settings and different socioeconomic conditions may affect the response of this service to climate change.

LAND TENURE AND OTHER FOREST RIGHT ISSUES

Deforestation and forest degradation in tropical and some of boreal forests are serious problems that contribute to the emission of greenhouse gases as well as to the fragmentation of forests. Deforestation and degradation have a series of direct and underlying causes, but none of these can be resolved if land and forest tenure are not clear or are not enforced. State land is more frequently subject to conversion into agricultural land than privately owned land. Privately owned and concession forests, however, are increasingly coming under pressure, especially in countries with policies that recognize traditional rights or favour the rights of community inhabitants to their surrounding forests. In the Amazon region, community lands also receive increased pressure, possibly due to the regional infrastructural plans (IIRSA), speculation of future forest values under new international agreements on 8 climate change (REDD+), investment in bioenergy, the relatively large size of many community lands in relation to their population and the lack of financial and human resources to secure their borders. Since many of the areas with land and forest rights concerns are in remote areas and refer to areas where people may have conflicting interests, regularizing these rights has been a major challenge in the past. Some progress has nevertheless been made in Latin America.

CHANGES IN POLICY ENVIRONMENT

REDD+ Expectations

Probably one of the more notable short-term changes in the policy arena is the discussion of GHG emissions reduction through REDD+ and management, conservation and restoration of forest carbon stocks. Large sums of money have been pledged against the demonstrable reduction of GHG emissions through REDD+, but so far, no international agreement has been reached on emissions reduction targets for developing countries. Further, in many pilot projects, measurable results have been interesting but financial benefits limited. REDD+ expectations are manifold, depending on the interest group. Some of these expectations are justified, others not, and most are probably too ambitious. Implementation of REDD+ strategies will have to deal with most, if not all, of the challenges mentioned in this chapter. At the same time it will require the implementation of a monitoring system, the extent and detail of which has not yet been agreed upon. While this has serious implications, the current (international and national) political environment is set to enable projects and countries alike, to meet at least some of these challenges. For the forest manager much of the challenge lies in adjusting management practices in favour of carbon accumulation, while at the same time maintaining biodiversity, recognizing the rights of indigenous people and contributing to local economic development.

Changes in Legislation

In Latin America, many countries implemented new forest legislation in the period between 1995 and 2000. While in some countries this was based on a thorough analysis of the forest sector, in others it was more in response to different pressure groups and based on changes in neighbouring countries. In some countries (for example Costa Rica), new legislation was relatively successful in achieving the objective of forest conservation , although reducing forest use for timber production considerably (Louman, in print). In others, it has been difficult to implement new legislation if unaccompanied by other measures and if the process was not participatory and consultative. More recently, countries have realized that they have better results when their new legislation is developed using more participative processes (for example in the DRC and Honduras). However, these processes are too young to be able to assess the true success in terms of increased implementation of legislative requirements.

Climate change will increase the challenge of designing and implementing new legislation that considers new international agreements, conflicts of interest in forest areas, as well as the need for coordination with other sectors. This may involve legislation on land and forest tenure, indigenous rights, the production of fuels and land use planning including restricting the access and use of certain areas or of some species, due to the risk of climate change impacts or the need of soil and water protection or maintenance of biological corridors. In revising forest legislation, it is important to consider all related legislation, so that, for example, legislation or policies oriented at increasing forest area on private land is not nullified by policies or legislation that define forest land as 'un-used' or 'luxury possessions', taxing them relatively heavily or even threatening to expropriate the owners.

Social Responsibility Requirements

Concerns for sustainable development, for the deterioration of the environment and of social relations, as well as for the negative effects of climate change at different scales are influencing market decisions. This can above all be noticed in agricultural product markets, where buyers are looking for products that meet specific environmental and/or social standards. Some banana plantation owners that export to the European market, for example, have started to invest in forest land for conservation and carbon emissions compensation.

New standards have just recently been developed to monitor and evaluate carbon dioxide equivalent emissions from livestock farms in Costa Rica, while the COOPEDOTA coffee cooperative was recently declared carbon neutral. These new developments pose interesting opportunities, more research is required to determine how these mechanisms can be used to improve the maintenance of other ecosystem services (such as water regulation and biodiversity maintenance), strengthen the adaptive capacity of natural and human systems and complement conservation and sustainable use of the existing forest areas within the agricultural landscapes.

Opportunity Costs of Land Use

Meeting REDD+ expectations has much to do with being able to identify the opportunity costs of local actors when they choose forest conservation and management rather than other land uses. Many of the REDD+ cost analyses are based on compensation for lost opportunities, although it has been found that forest conservation on private lands does not only occur for financial reasons. If lands surrounding forests have high opportunity costs, there is the likelihood of increased pressure to convert those forests to the adjacent land use in order to make them more profitable. Opportunity costs may vary due to variations in market prices of the crops cultivated, government policies that subsidize agricultural inputs or the exportation of the outputs, or policies favouring the production of biofuel. The forest user or owner does not easily influence these factors. As a group, in particular, if acting within the framework of REDD+, it may be possible to influence legislation, reduce the unequal treatment of forests as compared to agricultural crops, thus making forest management more competitive with other forms of land use.

UNCERTAINTY AND RISK MANAGEMENT IN FOREST CLIMATE

Climate change projections for the future involve a series of uncertainties. It is still not sure what emission scenario will best reflect reality, how these emissions change climate, in particular in relation to the distribution of precipitation or what other factors may play a role in influencing local vegetation and how local vegetation will react to climate and other factors. Thus, forest management for climate change has to deal with a range of uncertainties. The challenge is to reduce those uncertainties and to design management systems that can deal with unexpected changes. Uncertainty and risk management options may involve monitoring systems (*e.g.* climate, biodiversity, production, and social impacts), early warning systems, working groups that analyze the implications of data obtained through monitoring, mechanisms dealing with risk of income loss, appeal systems for unpopular decisions as well as free prior and informed consent of indigenous and local communities. Flexible adaptive management approaches need to be a part of any management strategy that involves risk and uncertainty. Such strategies will need to include a set of tools, rather than one specific approach, to be able to switch from one to another tool, depending on local conditions, changes in those conditions, and success of already applied tools.

CLIMATE CHANGE MITIGATION AND ADAPTATION IN FOREST MANAGEMENT

The potential effects and significance of climate change on the forest sector. These impacts have varying consequences and are dealt with differently by forest managers. The possible operational options available to forest managers for addressing climate change mitigation and adaptation are assessed. In addition, the extent to which these options are being applied by forest managers is discussed, with the help of case studies. These examples, although not necessarily due to climate change, give good indications of what managers perceive to be good solutions to potential future changing climatic conditions.

FOREST MONITORING

Forest monitoring is very useful to detect changes due to climate change, natural disturbances or human activities. It has become a requisite in the context of climate change mitigation in particular, in relation to deforestation and forest degradation. Due to the potential benefits that accurate carbon monitoring may bring within the REDD+ framework, monitoring has developed greatly over the past few years and requisites on accuracy and acceptability have increased.

For monitoring in general to be successful, it needs to have clear objectives, be as simple as possible, and benefit the people that invest time and/or money in it. However, many times the objectives may be clear, but the activities that are needed to meet those objectives may be vague. This may be due to lack of experience or lack of certainty on how climate will change and how this possible change will affect different components of the forest and forest management. A tendency exists to want to monitor everything that might possibly change, resulting in impractical and expensive monitoring proposals. As a result, monitoring for the impacts of climate change on the forests and people related to the forest is still just emerging. Monitoring helps us to identify changes and evaluate tendencies. Monitoring does not necessarily tell us the reason for these changes and tendencies, unless previous research has established such causal links. The next step would therefore be to analyse whether such changes correspond to changes in climate characteristics and then, to analyse whether such tendencies are negative or positive for the forest and the forest managers and whether actions can be taken to reduce the negative consequences and increase the positive ones. Current discussions on the implementation of REDD+ are occurring at the national level, however most of the monitoring experience has been obtained at the forest management unit level. While monitoring needs at these levels differ, they are highly complementary and any carbon monitoring system should consider linking these levels. It is very important to include all stakeholders to ensure agreement on the methodology and the variables to be monitored. The involvement of local actors has been shown to have two advantages; it is cheaper and creates greater ownership of the monitoring results.

In spite of the importance of monitoring for SFM and for preparation of responses to climate change, it still is not a common practice. Particularly in developing countries, few forest managers have the resources (human and financial) to implement these assessments.

Monitoring of Changes

Adaptation of forests requires in the first instance the identification of the changes that may occur and to which adaptation may be necessary or desirable. Although in general terms forest change scenarios can be developed based on global and regional climate change projections, the exact changes that will occur are not well known. There are several reasons for this uncertainty; the uncertainty in the climate change models in general, the scale at which climate change projections are made, the 12 inherent adaptive capacity of species and the communities they are in and the effect that interactions between species may have on adaptive capacity. In some areas, the changes that have been projected are drastic. The northeastern Amazon, for example, may lose most of its forest cover because of massive forest dieback due to droughts, giving rise to savannah vegetation. However, the rate of change and the exact result is not that clear. Other areas may follow suit at different rates and with different results. It will be difficult for forest managers to react to these changes, especially if it is not clear when and how these changes occur.

Adaptation strategies will need to include monitoring systems on climate, vegetation, fauna and essential non-biological components of the forests such as water availability. Without such monitoring systems, the forest manager will be grappling in the dark when making management decisions. In forestry, such monitoring systems are important, particularly because of the long time lapse between management actions and forest response. For this reason, permanent sample plots (PSP) are an integral part of SFM. Their main contribution to SFM has been a better understanding of the dynamics of forests and plantations. PSPs have been used for stock-taking (both at a national scale and in continuous forest inventories), for monitoring of changes in managed and unmanaged forests (*e.g.* in certified forests to monitor changes in species composition and structure), and for research purposes (*e.g.* the effect of silvicultural treatments and harvesting on species composition, structure and biodiversity). PSPs are less useful to measure changes in the diversity of fauna, impacts of forest operations such as harvesting and impacts on ecosystem services that go beyond the forest plot boundaries (for example water flow).

In countries that have long-standing experience with PSPs and a good network of meteorological stations, PSPs may provide a good contribution to the analysis of the effects of climate change on forests. While PSP are good instruments to detect changes at the stand level, forests are also influenced by changes that occur on a landscape level, *e.g.* water quality affected by sedimentation. To detect such changes, a combination of remote sensing techniques and a network of PSPs is probably the most appropriate strategy: remote sensing to detect changes in forest areas, and PSPs to detect changes in forest quality. Since remote sensing images and their interpretation for forest management is relatively costly for the forest manager, such monitoring is best carried out by organizations or associations that are responsible for larger areas or a group of stakeholders. This will require, that all potential users of the monitoring information agree on a common set of variables that are useful for forest management decisions and should therefore be monitored. An important part of monitoring systems is the database and processing of the data. This usually requires major investments in human resources but some companies have been able to develop their own computer hard and software that allows for quick data storage and analysis.

Monitoring Techniques of Animals

The techniques used monitor animal populations, particularly the larger mammals, depend on the objective of sampling. More local research is necessary to identify the techniques and variables to be sampled in specific cases (*e.g.* for a particular species in a defined region). Climate change can shorten the life cycle of insects, increasing their reproduction rate and the risk of infection and damage. Traps in sampling points can help to detect rapid increases in population sizes. The traps will need to be specific for the insects to be monitored, have appropriate bait and be placed at the right position in the forest. Turchin and Odendaal (1996), found that one funnel trap, used to trap southern pine beetles in the united States, were good for covering an approximate area of 0.1 ha. Some insects are more ground related (*e.g.* dung beetles) while others (*e.g.* butterfly families) may fly in open or closed forest areas. Although these latter insect groups have not been related to pests, they have been successfully used to identify changes in forest structure and composition related to fragmentation and tree harvesting, and may be useful to detect forest changes due to climate change. Specific dung beetles may be related to specific mammals and butterflies have been related to openness of the forest and may be an indication of dieback. Further research is needed to fully understand these relationships. Larger animals may be trapped (as in the case of small rodents), or counted visually using walking transects. Animal tracks may also be used as an indication of the presence and abundance of species. However, care should be taken that sampling density is sufficient to formulate robust conclusions. Steele *et al.*, (1984) concluded that three repetitions of a 2 km transect was sufficient to determine species abundance, richness and diversity of large animals, but it was more difficult to estimate small mammal richness and diversity. In general, design of a monitoring system requires expert knowledge, but local communities can be trained as parataxonomists to implement the monitoring.

Due to the need for additional information on species behaviour and preferences, as well as the relatively high time investments needed for animal monitoring, it is important to identify, as early as possible, those animals (and plant species alike) that are more susceptible to climate variations. Abundant animals may be easier to monitor, but many of them may also be less susceptible to changes in climate and the environment. Usually the species with a small range and short generation time are more responsive.

Forest Fire Monitoring

Monitoring of forest fires contributes to our knowledge on the extent of deforestation and forest degradation. Such monitoring is traditionally done through patrolling forest areas and operating watchtowers. The development of remote sensing techniques has made it possible to come to ever more accurate and timely information on forest fires, above all in large uninhabited areas. Laneve *et al.*, (2006) estimate that if images can be obtained at a sufficient spatial resolution to detect 1500 m^2 fires at 30 minute time intervals, this will be sufficient to reduce the number of large fires in the Mediterranean forest of Italy. For small forest land holders and many communities, such technology is not available and even the construction of towers may be too high an investment. Patrolling, however, has shown to be an effective way of forest fire prevention in community forests in Guatemala. During the last decade several proposals have been made to set up fire detection systems using wireless sensors , but most of these have not emerged from the experimental phase, possibly due to costs and the problem of maintaining the network.

CAPACITY OF FORESTS TO RESPOND TO CLIMATE CHANGE

The adaptive capacity of forests, for the purpose of this document, is understood to be the inherent ability of the forest to adjust to changing conditions, moderating harms and taking advantage of opportunities (Locatelli *et al.*, 2010). Thus, strengthening of the adaptive capacity is oriented at increasing the resistance or resilience to changes but may also include adapting the forest to new conditions by facilitating changes in the system (*e.g.* by species introduction). In general, strengthening the adaptive capacity of forests aims to maintain, restore or enhance forest area, biodiversity and forest health and vitality. Many of the actions oriented towards mitigation of climate change through REDD+ have a strong potential for synergies with actions oriented at strengthening the adaptive capacity of forests, in particular if such actions consider ecological safeguards, such as biodiversity conservation. Experiences in strengthening the adaptive capacity of forests to climate change have been more widespread in plantations and agroforestry systems. These systems tend to have a simpler structure and composition that makes it easier to detect changes due to climate change and to design and implement adaptation-strengthening mechanisms. This is much more difficult in complex natural forests, in particular in the tropics. However, because of their simplicity, these systems may also be more vulnerable and therefore the need to look at adaptation options is greater. Interestingly, several of these adaptation activities are oriented towards making these (agro) ecosystems more diverse.

Maintaining Forest Area

Larger forests usually have greater species diversity and cover a greater variety of sites, thus reducing the risk of losing the whole system if climate change negatively affects several species or specific site conditions. Forest management appears to be the solution; both wellmanaged protected areas and well-managed community and private forest concessions in Guatemala and the South of Mexico have shown to be more effective in avoiding deforestation, fires and forest degradation than areas poorly managed areas. This probably goes beyond purely economic considerations.

Land and forest tenure, recognizing the benefit of maintaining forests and joining forces with other forest managers are all important requisites. Size of forest area also seems to be important, for both individual landowners and community or multiple owners. Ecologically, larger forest areas show less edge effects, while from the management point of view, larger sizes allow for economies of scale.

Managing natural forests often is recognized as a claim on that forest. If good relations are held with local people, such claims are well respected. Owning forest but not managing it, has often resulted in unauthorized entry by third parties for the extraction of timber and NWFPs, or for conversion to agricultural land. Managing the forest but not entertaining good relations with the neighbours has often resulted in forest use conflicts, at times ending up in armed conflicts or burning of parts of the forest estate.

Such relations are more important in large forest tracks, since in these it is harder to establish continuous human occupancy. Good forest management normally includes fire, pest and disease management. Of these, fire management may be the most significant in maintaining the forest area, although serious pests, such as the mountain pine beetle in pine forests in North America, may also contribute to substantial forest loss. Managing the forest may be costly and income from the sale of one or more of its products may not off-set the extra cost of management. However, often, costbenefit analyses compare conventional operations (without much strategic planning or considerations for biodiversity or forest dependent communities) with managed operations.

From a private forest owner's point of view, this may be reasonable, but in practice, this approach has been used to justify continued conventional harvesting operations, giving the forest sector a poor image and increasing the pressure on governments to impose stricter regulations. In countries with greater willingness of the private sector to participate in 15 improving forest management, a series of alternatives were found to either make forest management attractive or propose other forest-based income solutions. Usually this was done by a carrot and stick approach: if a forest manager does better than the legislation requires, they receive subsidies, discounts on taxes and are a preferred provider of specific ecosystem services. While these approaches in theory seem to be very promising, in practice they have not had the expected outcomes. This is partially because of the high financial and administrative cost to actually obtain the carrots and because forest owners and managers were not aware of the existing opportunities.

Management of forests, therefore, should go beyond the mere planning of protective or productive activities and not depend on one single form of financial income. Forest managers need to be informed and aware of local, national and international opportunities for income generation. New opportunities may be through tax or fee discounts for good forest management (as the case of harvesting fee discounts for certified forests in Peru), payment for environmental services (*e.g.* Costa Rica and Mexico) or niche markets (*e.g.* markets for specific NWFPs and carbon).

Due to the relatively limited demand of these above mentioned products and services and with only a few exceptions (*e.g.* Brazil nut gathering in Bolivia, Stoian, 2004), none of these have been able to single-handedly pay for management and protection of large natural forest tracts. Only where combinations of products and services were obtained has management of natural forests become a serious land use competitor. New opportunities may also work for forest plantations, where payment for environmental services, such as carbon sequestration, may at least partially off-set the initial establishment costs. However, selling sequestered carbon at the end of the rotation has little effect on the overall profitability of plantations, due to the low carbon price and high interest rates. The present value of these future sales is very low and rarely will change a non-profitable exercise into a profitable one. Selling less carbon earlier on during the rotation, or shortening rotation length, are two options that may make timber plantations more attractive. Establishing tree plantations not for timber, but only for carbon or other environmental services, as yet has to show its profitability for the forest owner or manager and is usually only accomplished where the forest owner or manager also garners important non-tangible benefits from those forest.

Conserving Biodiversity

Maintaining forest area is of course a good means to maintain a certain level of biodiversity. However, as can be seen from some countries, increasing forest area does not necessarily increase biodiversity nor does it necessarily mean that old forests or undisturbed forests are maintained. In many countries, reduction of net deforestation figures is at least partially due to compensatory measures, such as natural regeneration and plantations. Although depending on how these new forests are being managed, and how close they are (in time and space) to the original natural forests, these usually do not have the same species composition and biodiversity as the lost natural forests. In terms of capacity to adapt to climate change, the change in species composition may sometimes be an advantage, if new species are better adapted to changing conditions. More problematic may be loss of diversity. Loss of diversity will make forests more vulnerable to changes, since they will not have the rich gene and species pool from which to select for the new conditions. In this respect, care should be taken of the trade-offs between mitigation and adaptation objectives; too great an emphasis on management for carbon may reduce structural and compositional diversity, thus reducing the system's inherent adaptive capacity.

A noted change in forest management in the light of climate change has therefore been an increased interest in maintaining or increasing diversity of the forests. Mixed species plantations, use of a larger number of clones and reductions in the scale of harvesting operations have been implemented as measures to maintain or increase biological diversity. These same measures are now receiving more attention because of their potential benefit in preparing forests for climate change. In addition, literature suggests that the potential yield increase from 16 appropriately selected species mixes will more than outweigh the additional costs that may be involved in mixed tree plantation establishment. The use of nitrogen fixing tree species as part of the mix, in particular in degraded lands, is beneficial for overall growth rates.Reducing the scale of harvesting operations is one way of increasing the possibility of ecological connectivity between forest patches. Plantation establishment is also an important measure that may achieve this as does planting of trees outside the forest.

Although some practices are being adopted and theoretically will contribute to maintaining biodiversity, there is still a need for further research. For example, we do not yet know how much biodiversity change will cause a major and irreversible change of forest types or even ecosystems. Some authors suggest that maintaining functional diversity may be sufficient. However, it is unclear how susceptible diversity will be to climate change, if, for the different functions, the number of species that provide diversity function is strongly reduced. It is also unclear how species will react to climate change, by themselves, in combination with other species and in combination with a number of environmental and human factors. Continuous monitoring of the forests is critical to providing insight into these interactions.

Maintaining Forest Health and Vitality

The main threats to health and vitality are pests, diseases, fires and extreme weather events. In addition, diversity usually strengthens health and vitality and therefore the actions mentioned above to maintain or enhance biodiversity, also are useful for maintaining health and vitality. A number of silvicultural techniques have been developed for maintaining health and vigour of a stand. Removing old, poorly formed and damaged trees, for example, reduces the risk of spreading diseases and pests, although at the same time it may reduce diversity and thus increase the susceptibility of forests to diseases and pests. Applying such treatments requires knowledge of the specific risks related to individual tree species and the potential benefits of maintaining poorly formed trees in the forest. Using harvesting residues on the forest floor may increase availability of nutrients for the remaining trees, thus increasing vigour, but may add to the fuel load and therefore increase fire risk. Nevertheless, timing of use (beginning of wet season) may increase the benefits and reduce the risks. In plantation forests, a reduction of old growth and an increase in the relative presence of young stands, enhances the general health and vigour of the forest from the point of view of timber production in the medium term. Again, however, it reduces diversity, thereby reducing the adaptive capacity of the forest to externally driven changes. The decision to reduce such growth in favour of young stands needs to be taken in consideration with local conditions and management objectives.

Reducing Risk and Intensity of Damage

Reducing the risk and intensity of pests, diseases, fires and hurricane damage, along with managing the hydrological cycle, will become major concerns for many, if not all, forest managers under changing climatic conditions. Due to the complexity of measures that may have contradictory effects, there is the tendency for integrated management practices; for example, combining insect control with monitoring exercises and implementation of management practices that reduce susceptibility of the forest to insect attacks. Such practices include those treatments that help maintain the vitality of the forest, including timely thinning and species mix. For most regions however, few comprehensive management plans exist, and in most cases, plans emphasize monitoring and combating pests and diseases, rather than preventing them.

The example of pine beetle management in Central America shows how management oriented towards the prevention of fires may also reduce the risk of insect pests. Integrated fire management in forests has been promoted in many countries and by international and national agencies alike. The proceedings of a 2009 seminar in California summarize many of the experiences and achievements to date. Martell (2007) defines forest fire management as: "getting the right amount of fire to the right place at the right time at the right cost". Integrated fire management includes the following components; prediction of fire occurrence, fire prevention, fire detection, initial attack, fire management, strategic planning of resources, and fuel management. Training, knowledge sharing and planning of the different components of fire management is extremely important.

Prediction of fire occurrence requires the identification of the main causes for forest fires, followed by an analysis of factors that influence the frequency of those causes. In Mexico, this has been done through proxy indicators, such as nearness of populations and types of land use (danger assessment). This needs to be combined with characteristics of the fuel, temperatures and moisture content (risk assessment) as well as with an assessment of the value of the resource in order to be able to set priorities in fire prevention and combat activities. In many developed countries, this information is combined with daily weather forecasts to assess fire danger ratings. This has proved to be a useful early warning system.

Fire prevention is probably the most cost effective and efficient way to reduce fire damage. It requires identification of frequency, size, severity and causes of human induced fires and needs to be followed by awareness campaigns to reduce those causes or projects oriented at training and introduction of alternatives to burning forest and grazing lands. In many regions, for example Mediterranean countries and western states of the United States, watch towers are the main tool for fire detection. Patrol flights are used as well but it is difficult to plan them cost effectively. Once detected communication of location, size and burning characteristics of the fires among the fire fighters and to the public is very important, especially for the initial attack. Whereas in North America it is a problem of deciding how many air tankers to keep on standby for the initial response to fires, in many tropical countries air tankers are not available and initial attack may need to be done by local fire brigades. The quicker the response time, the smaller will be the risk of uncontrolled fires. However, such fire brigades require an extensive programme of training of the local teams, as well as sufficient equipment to be able to fight small fires. Initial attacks become more efficient if these are based on previously prepared plans that consider potential impact of fire fighting, as well as the potential risks of damage to natural, human and physical resources.

Fire management requires continuous monitoring of fuel conditions, fire behaviour and climate. With this information and short-term predictions on their changes, fire fighters can decide when and how many people to dispatch to control the fire. Fuel management is one of the more common and discussed approaches to forest fire reduction. Protection of forests has led in many cases to the accumulation of fuel, increasing the risk of forest loss, rather than reducing it. Prescribed burning is a common tool, but this needs to be supervised by experts, in order to avoid the fires to get out of control. Cutting of fire breaks is another measure often taken. These do not only form barriers to fire progress, but may also help accessibility of the forest for fire fighting crews. Technology has proved vital in monitoring and control of forest fires in regions such as North America and Europe. However, in tropical countries, where access to this technology and to the forests is more limited, fire management strategies need to include all stakeholders and all related sectors since most fires originate outside forest lands and involve non-forest stakeholders.

Risk reduction strategies for damage from hurricanes or cyclones need to consider damage to forest, agricultural crops, communities and infrastructure. Reducing risk of damage to forests can be done through silvicultural practices such as shorter rotation cycles; young trees often are more resistant to wind throw, but if thrown, will result in less biomass lost compared to larger trees. Increasing resilience, for example by maintaining good seed sources and mixed species forests, including species that readily sprout after wind throw is another means of reducing damage. In Toncontín, Honduras, the local community started to improve forest management in 1997, one year before hurricane Mitch struck. The forest is dominated by species resilient to hurricanes through different strategies including rapid and abundant seed regeneration or re-sprouting. Part of the management improvement was the retention of seed trees. Comparative studies one year after Hurricane Mitch indicated that the forest with seed tree retention was recovering its original species composition faster than forests where all commercial trees had been cut without seed tree retention.

Improving Water Regulation

Changes in precipitation patterns probably will affect forests more than changes in temperature at least over the next twenty years. The impacts will depend on the increase in atmospheric CO2 and the reaction of trees to this increase. The potential water regulation functions of forests therefore, are receiving more and more attention in the forests and climate change discussions. The role of forests in this case is vital not only as a provider of an ecosystem service (in particular in relation to the quality of water for consumption, for irrigation, for industrial use), but also for the survival of the forests themselves. Many myths regarding the function of forests in the water cycle exist. Much more research is needed and this research should include studies on the influence of individual trees and forest stands on the local water cycle and the water cycle of watersheds. As a rule of thumb, the original natural vegetation is the best way to regulate the water cycle. Planting trees in grassland swamps, such as one of the water catchment areas of Bogotá, Colombia, may result in rapid depletion of the water resources. On the other hand, leaving trees in dry areas may improve infiltration, soil structure and water holding capacity of the soils. Planting fast growing exotics in such areas may result in depletion of the soils. Dry areas, however, may also be affected by salination and in that case, care should be taken not to increase underground water levels, since this may increase the release of minerals in the upper soil layers.

Different plants have different water use efficiencies and these may change when exposed to more sunlight. Oak in Italy appears to be less sensitive to light conditions in its water use efficiency than a local Beech species. This may imply that beech stands may suffer less from reduced rainfall, since individual trees will react positively to lower stand densities while oak stands, on the other hand, may lose volume. In terms of effect on the regulation of the water cycle, oak may be easier to manage, since thinning of the oak stand will reduce water use more than in beech stands. Couralet *et al.*, (2010) found similar results for species in understory trees in Africa. Equal treatment of a forest stand in terms of modifications to the microclimate and water availability will have varying effects on the different species. This relationship, however, needs more research. Knowledge of this correlation is useful for adaptation, species selection for local conditions and the estimation of the value of a species and its management system for mitigation. In Argentina and Chile, Nothofagus pumilio is suitable for mitigation due to its plasticity, while the regeneration capacity of N. macrocarpa was highly affected by the moisture gradient within canopy openings. Dietz *et al.*, (2006) in a study in Indonesia, found that tall trees may increase water evapotranspiration and decrease throughfall. This suggests that smaller trees would be better conservers of water. Of course, this is dependent on how well these smaller trees allow the throughfall and stem flow water to infiltrate in the soils, and the water retention capacity of the soils.

Apart from the effects on plant growth of individual species, climate change is likely to affect water availability at a watershed level. There has been an increase in the implementation of projects in upper watersheds with many positive benefits. However, the meta-analysis of Locatelli and Vignola (2009) clearly indicate that this is not always the case. Two strategies can be followed in such cases. They demonstrated the necessity to study the effect of trees on an experimental basis, before promoting large scale regeneration or plantations. In addition, 20 monitoring of water flow at strategic points within the watershed, taking into account current land use practices will allow for changes in the planning of species composition. In most cases, at the landscape level, control of land use and forest regeneration goes beyond the control of individuals and will need some form of stakeholder collaboration to achieve the desired results. A good case of stakeholder collaboration can be seen in the Dominican Republic, where farmers with land suitable for irrigation allowed owners of land on the ridges to occupy part of their land in return for reforestation on the ridges. It is estimated that for each hectare of irrigated land, thirty hectares are being regenerated, contributing to the availability of water for irrigation.

DEVELOPMENT MECHANISM AND OTHER CARBON INITIATIVES

As early as 1995 the European Union set up the European Trading System (EU ETS), which puts a cap on the country emissions but allows EU countries to trade with other countries whose emissions remain well below the cap. Then in 1997, the Kyoto Protocol (KP) was adopted, although it did not enter into force until 2005. As of September 2011, 191 states have signed and ratified the protocol. Under the KP, 37 countries committed themselves to reducing their emissions to a level below that of 1990 and all member countries gave general commitments. They also agreed that countries that have not been able to meet their quota may compensate part of that by either setting up 'joint implementation' projects with other countries, or invest in Clean Development Mechanism (CDM) projects in developing countries. Of the latter, afforestation and reforestation are the only forms of forestry projects eligible. None of the tree or forest carbon offset projects however, are formally recognized within the framework of the ETS. Thus, options for forest carbon trade are limited to the voluntary markets and over the counter trade. In addition, funding is available to finance the costs of developing countries to prepare themselves for Reduced Emissions from Deforestation, Forest Degradation, forest conservation and sustainable forest management (REDD+). The delivery of funds to the countries has however, not reached expectations. Only around 1.5% of the CDM projects, for example, are forestry projects, while at the same time, prices on the voluntary markets fluctuate with expectations of future demand but are nevertheless very low.

CDM Projects

Most carbon initiatives are either related to the CDM scheme or set up in a similar manner. These projects, with a few exceptions, are oriented toward sequestration of carbon. Trees are planted and a commitment is made to maintain the area under tree plantations for a minimum period of time, commonly 30 years. Although within forestry circles this is thought to be a reasonable time period, it should be noted that if after thirty years the trees are cut and not replanted, in the long term no carbon has been sequestered, other than that used in long term products (permanence). In addition, the capacity of most tree species to sequester is relatively low. Official IPCC figures, range from around 1 tC/ha/yr to about 10 tC/ha/yr. Within the scheme, requirements for carbon accounting are very strict, raising the costs of entry and thus limiting the participation of small holders in the scheme. In voluntary markets, carbon accounting standards may vary according to the expectations of the buyer. However, buyers are increasingly requesting the strict carbon accounting methods approved by CDM, sometimes accompanied by standards that measure the social and economic impacts of these initiatives.

REDD+

REDD+ is an international mechanism aimed at avoiding deforestation and forest degradation, promoting sustainable forest management and conservation or enhancement of the forest (carbon stock). The innovative part of REDD+ is the broad international attention and the funding potentially available for the resulting strategies, the focus on carbon rather than on wood and NWFPs, and the recognition that the solution lies in an inter-sectoral approach rather than only within the forest sector. Because of the expectations REDD+ has raised, there are many actors involved, some of them seeing REDD+ as a means to strengthen their own agenda, developed over years or even decades. Foresters see REDD+ as a means to bring forestry for timber management back on the foreground while conservationists focus on opportunities for protected areas. Many NGOs hope to be able to strengthen community forestry, while some indigenous people's organizations view the negotiations on REDD+ as a chance to finally receive formal recognition of their traditional rights. Consultants and investment agencies consider it as an opportunity to sell their services, and some of these may actually exaggerate the potential benefits in order to convince actors to participate. Fortunately, these agendas, in many cases, are complementary and a number of very interesting experiences exist, although not necessarily fitting into the framework of the international negotiations (which are oriented towards national implementation rather than project implementation) or national policies. A good example is the case of Bolivia, where the Noel Kempff project made promising advances in implementation but the lack of support of the national government has prevented further progress.

Many of the activities proposed under the REDD+ mechanism are not new but may be differently focused. The aim is to reduce emissions rather than produce specific products or conserve biodiversity. In addition, the approach towards implementation is different with more attention being paid towards processes, such as meaningful participation, free, prior and informed consent (FPIC), conflict resolution mechanisms and monitoring and verification. These are all related to decision-making and stakeholder consultation processes. Examples of projects include SFM initiatives, protected areas systems that have contributed to stall the advances of the agricultural frontier, restoration of degraded forests and reduction of fragmentation. These projects have all been successful in showing the contribution of these activities to avoided deforestation and forest degradation. In addition, successful projects have clarified land, forest and carbon rights, recognizing legitimate claims of the different stakeholders, establishing acceptable cost and benefits sharing mechanisms and implementing activities to avoid negative social and environmental impacts.

Very few of the projects have been designed to fit into national REDD+ strategies. However, lessons can be learned from the existing project experiences for implementation of the future strategies. Forest managers will most likely have to deal less with eliminating the underlying causes of deforestation and degradation, since many of those relate to policies, markets, culture and demographic factors such as population growth and migration. Within a national strategy, it is also less likely that they will have to deal with leakage since this is better dealt with at a broader scale (*e.g.* directed settlement areas, off-farm employment opportunities, legislation that prohibits land use change from forest to agricultural lands). Monitoring, respect for other people's rights, FPIC and activities to avoid negative social and environmental impacts will need to be designed to be effective locally and at the same time fit into the national REDD+ policy framework. The major tasks for the forest manager within a national REDD+ framework will be management and monitoring of the carbon stock. Managers will be motivated to do so if they perceive benefits from it, usually in the form of extra income and increased access to technical assistance and technology. 22 Management and monitoring of the carbon stock in theory should involve all five carbon pools; above ground biomass, dead wood, litter, underground biomass and carbon in the soil. Currently, for REDD+, the above ground biomass is the main pool being discussed, although for projects, the carbon accounting requirements may differ according to the standards used. The verified carbon standard (VCS), for example, considers different methodologies for different REDD+ type of activities. For reducing emissions from mosaic deforestation and degradation (with forest patches <1000 ha surrounded by agricultural fields), VCS include above ground tree-biomass, below ground biomass, dead wood and carbon in products in the carbon accounting. The actual accounting methods usually have to follow the IPCC 2006 guidelines.

The other carbon pools may change, but the direction of their change is less well known and depends on local conditions and future land uses. If the changes are expected to be positive, it is accepted practice to exclude the carbon pool, since this would lead to conservative estimates of carbon benefits. Soil carbon pools become more important in boreal and temperate forest areas and in non-forest land uses. In those cases, carbon stock in the soils is usually greater than in the above ground vegetation, and can be highly influenced by land use practices, water run-off and erosion. In project accounting, several pools are usually considered for the calculation of the baseline. Most of the current carbon initiatives concentrate on monitoring of the above ground biomass as a proxy for carbon content in above and below-ground biomass to provide inputs for accounting of the changes in carbon during the project life. Generally, such monitoring is done through a combination of satellite images and on the ground measurements. In some projects, communities are heavily involved in monitoring their own carbon stocks, thus promoting their appropriation of carbon management. For carbon management, the most interesting conservation sites would be those under highest pressure of conversion. In practice, however, REDD+ pilot initiatives are in areas under little pressure since this usually reduces the initial investments in terms of financial and human resources.

MARKET INFLUENCES ON ADAPTATION AND MITIGATION PRACTICES IN FOREST MANAGEMENT

In many places, forests have disappeared due to conversion to lands for the cultivation of high value crops. In other places, fall in prices of crops or beef have led to abandoned agricultural lands, on which forests regenerate. Increased prices of agricultural crops lead to increased production and increased need for packaging material, which in turn stimulated the establishment of plantations of fast growing tree species. An increase in the value of forest products and services may make it more attractive for forest owners and users to manage the forest than to convert them into marginally producing agricultural or livestock lands. An increased demand for international enterprises to behave in an environmentally responsible manner makes them incorporate forests and trees into their agricultural production plans. All these examples show that markets, whether for agricultural products, forest products or ecosystem services, have a complex and close relation to the way forests are used or conserved and thus can influence the mitigation and adaptive capacities of these forests. With the recent discussions on climate change, three types of relations between markets and forests have become crucial; markets for forest carbon, markets that require social and environmental responsibility of a wide range of companies and markets that influence the relative value of forest use in relation to other land uses.

MARKETS FOR FOREST CARBON

Since the KP has been in force, several carbon emissions trading schemes have been established. By far the biggest is the EU ETS, which currently trades more than 80% of all carbon credits with prices significantly higher than in the voluntary carbon markets. Within the scheme, large GHG emitters have been assigned a maximum amount of GHG that they may emit during an established period of time. Those that emit less than this 'cap' can trade the difference with those that emit more, earn emission reduction units through the Joint Implementation programme of the KP flexibility mechanism, or buy Certified Emission Reductions (CERs) through the CDM, also part of the Kyoto flexibility mechanisms. The EU ETS does not recognize sinks, effectively eliminating forest carbon credits from the scheme. Forest carbon credits can be traded in the voluntary carbon market and its proportion in that market has increased considerably after REDD+ was recognized by the UNFCCC as a regitimate form of emission reduction. Overall, however, the value of carbon traded in the voluntary market has declined, possibly because the expected international agreements did not get through, nor did climate change legislation in the United States.

The World Bank (2011) indicates that future demand and supply of carbon are expected to be in balance and it is not very attractive to set up new carbon initiatives, unless GHG emission reduction commitments in developed countries are strengthened. It is, however, difficult to predict the demand, since it depends also on other (macro) economic factors, such as fossil fuel prices, type of economic development and growth. There is another way of trading carbon, through direct transactions between buyers and suppliers. Some examples exist where communities or community associations in developing countries entered into agreements with companies in developed countries. Such companies are often prepared to pay more than market prices for carbon credits because they also look for additional benefits, usually contributing to their social and environmental responsibility.

Social Responsibility Requirements

With the focus on conservation and management of the carbon stock, other types of markets are being reconsidered as well. These are markets that promote sustainable forest management and also favour climate mitigation and adaptation. Forest certification, for example, requires a number of the same forest management and conservation requirements as does REDD+, *e.g.* no deforestation, limited degradation by reduced impact logging practices and fire and pest management.

Several forest certification schemes exist, but globally the two best recognized schemes are the Programme for Endorsement of Forest Certification (PEFC) and the Forest Stewardship Council (FSC). The former endorses national forest management standards, includes the verification of origin of the timber products through a chain of custody certification and already covers more than 200 million hectares of forests, but does not have as good a standing in the market place as FSC. FSC only endorses national indicators that are linked to an international standard and that have been agreed upon through a national process during which there has been equitable participation of social, economic and environmental stakeholder groups. These two certification schemes differ essentially in strictness of social and environmental requirements, appeal procedures and stakeholder participation. FSC is stricter and has more elaborate participatory processes than PEFC while PEFC has greater independence between standard setting and certification bodies. Both schemes are adjusting and are beginning to resemble each other. Forest certification in general, however, only applies to specific forest management units and therefore has a limited effect on overall deforestation or reforestation rates. These positive effects will depend very much on the scale and intensity of the operations.

A greater positive effect may be perceived on forest degradation. Since certification requires the application of reduced impact logging techniques, it has the potential to reduce damage due to felling and extraction to about half of that of conventional operations. Depending on the scale and intensity of the operations this may imply a considerable reduction in emissions. Increasing processing efficiency of the timber, once it has been logged, has a further potential to lower emissions. FSC certification has the potential to contribute to the reduction of emissions from forest degradation, since it has a performance based system.

Although forest certification was originally set up to increase market benefits - access and price - especially for tropical forest producers, currently only market access has been realized. Price increases have been limited to certain cases or in size. Easier access to credit and technical assistance has been seen as an additional benefit. All of these are factors that help increase the adaptive capacity of the forest manager; better organization, greater or more stable income and better access to supporting services.

Forest certification also has the potential to also increase the adaptive capacity of forest dependent people; additional training allows them to diversify their 26 economic activities, new employment opportunities help increase and diversify their income and local organizations involved in forest management are strengthened which enables them react to sudden environmental changes. An example of the latter is the village of Toncontin, in Honduras. This was the first village that was able to distribute emergency supplies after Hurricane Mitch hit in October 1998 because they could quickly identify potential beneficiaries and set up a distribution system based on the existing forest management organization.

Other certification schemes, such as schemes that certify biological agriculture or good agricultural practices, also have had a positive effect on the forest, both in area and in quality, and thus have the potential to contribute to both forest mitigation and forest adaptation activities. In Costa Rica, banana companies have purchased forest areas for protection and restoration to offset part of their emissions in the production process and contribute to biodiversity conservation.

Throughout the tropics, selected coffee plantations are increasing tree cover to be able to sell their products as 'shade grown' and more recently efforts are on the way to estimate GHG emissions from coffee production to be able to determine the level of carbon neutrality achieved. Carbon sequestration in trees is often seen as a good option to offset some or most of the unavoidable emissions. Such low emission schemes are very promising considering that they may contribute to carbon sequestration and storage -

agroforestry systems with coffee or cacao have been reported to store as much carbon as 20 year old secondary vegetation - as well as to strengthening the adaptation capacity of the local agro-ecosystems and their inhabitants, contribute to biodiversity, water regulation and disease and pest control if the shade is well managed. These schemes may also contribute to reducing the level of risk producers face when diversifying their production and securing more stable markets for their cash crops. They may be particularly useful in the context of landscape management, where they offer the opportunity to strengthen management of the ecosystem services throughout the landscape, including in agricultural fields.

Another social responsibility scheme oriented at sharing costs and benefits of ecosystem service management, is the concept of PES. Such schemes may not necessarily be oriented towards carbon sequestration and storage but nearly all existing PES schemes promote the maintenance or enhancement of forest and tree cover. They may or may not require the measurement of the services provided. Some assume that if forest cover is maintained or expanded, or its use limited to carefully management operations (timber, NWFP, conservation, tourism), the desired services are provided. Such is the case of the Costa Rican PES scheme, which pays a flat rate per hectare per year for a mix of carbon sequestration and storage, water regulation, biodiversity maintenance and scenic beauty for each hectare of forest. In Mexico, the first three services are being paid for through separate PES schemes, although all three schemes may have similar results in terms of carbon sequestration and storage, water regulation and biodiversity maintenance. In particular where water resources and biodiversity are at risk due to climate change, the contribution of PES to adaptation may become important. In general, such schemes also offer additional income to producers and forest owners, but in each case the benefits should be evaluated against the costs of meeting entry and monitoring requirements.

Managing Uncertainty and Risk

The consideration of risk has always been an important part of forest management activities. With climate change, uncertainty is increasing as are the risks of long term investments in forestry. Below are two examples of how the incorporation of uncertainty and risk into management decisions and improved monitoring for early warnings of changing conditions.

The Birris Micro Watershed

The Birris micro-watershed is located in Costa Rica. Its main landuses are cattle farming (milk and meat) and horticultural activities, while the micro-watershed provides a small hydro-electric power plant with water, later flowing into the Reventazón river with a larger hydro-electric power plant. The micro watershed is likely to suffer from climate change through changes in the onset of the growing season, as well as more severe droughts and high intensity rains. Erosion and sedimentation are likely to be the main direct impact of such changes and are considered by the different stakeholders as being a major issue: for the farmers because they lose fertile soil; by the electric power plants, because it fills up the reservoirs and may damage the machinery. These potential problems were analyzed by different stakeholders and indicators were developed to measure the potential effects of different land use scenarios on components potentially important for climate change adaptation and mitigation capacity of the micro-watershed. Four scenarios were identified by their potential to reduce sedimentation;

- business as usual,
- all critical areas reforested,
- all critical areas under improved soil management and,
- a combination of good soil management throughout the micro-watershed and reforestation in some of the critical areas.

Scientists estimated the values of the indicators for each of the four scenarios and once presented with the result of these studies, participants were requested to agree on one of the four land use options. They chose for an intermediate land use option that requires low initial investments while still reducing the risk that climate change may cause much erosion, and degradation of the agricultural lands and sedimentation of the hydro-electric power dam. Development objectives were thus combined with an assessment of risk of negative impacts of climate change, in this case erosion due to increased rainfall intensity.

Indicators of Socio-economic Impact of Land Use

Part of the research undertaken as part of a Mesoamerican Agro-environmental Programme (MAP) funded project is focused on identifying indicators that could predict the consequences of land use change, irrespective of the origin of the change. Studies have been carried out on the socio-economic and environmental impacts of different land uses in some Latin-American countries (*e.g.* Brazil, Costa Rica, Nicaragua) and this information was compared with information on past land use changes in those areas. This analysis should give rise to indicators on the timing and drivers of land use change as well as the potential consequences of these changes. The concept is very similar to the Pressure-State-Response set of indicators used by many countries at the national level within the framework of the UN sustainable development monitoring. The difference between the two lies in the scale and detail, with the MAP project allowing for the downscale of monitoring to landscape level, giving local authorities and stakeholder platforms the tools to predict change and estimate the risk of the negative impacts of those changes. This helps the operationalization of national sustainable development and climate change monitoring systems.

INCREASE ADAPTIVE CAPACITY OF ECOSYSTEMS THROUGH FOREST MANAGEMENT

Forests are important in increasing the adaptive capacity of forest communities, in particular as a safety net in case of emergencies and as a form of diversification of economic activities where NWFPs can be harvested in a sustainable manner. Below, some specific examples of forest management activities that contribute to strengthening local adaptation strategies are discussed.

Management of Tree Cover to Regulate Water Availability

The importance of forest management for water regulation cannot be overstressed, since water resources are likely to be the resources most affected by future climate change in many parts of the world. Tree cover can be managed to regulate water flow throughout the year, to increase or reduce water flow, or to protect river and coastlines against erosion at times of high water levels. In some cases, such management is promoted through PES schemes, allowing for a more equitable distribution of costs and benefits. In other cases, conservation efforts are oriented at improving the livelihoods of people living in the buffer zones, thus reducing their dependence on the forest for daily subsistence.

Management of Hunting

Bushmeat is one of the major NWFPs and in cases of emergency, may become the main source of food for local people. Maintaining viable populations of animal species therefore allows an important emergency supply to be maintained. The main threat to animal populations is hunting beyond sustainable levels. This may occur because of commercial hunting, natural population growth and migration.

Over-hunting may lead to unexpected changes in other parts of the forest or its ecological processes. A well-known case of unexpected effects is known from Yellowstone Park in the United States where wolves had been hunted before the park was declared a protected area. This resulted in uncontrolled deer populations within the park, leading to overgrazing and the lack of regeneration for a number of tree species. Hunting of Agouti in Brazil, combined with the elimination of their preferred forest habitat, has led to a decrease in the of regeneration of the protected Brazil nut tree, since the Agouti is the only animal species capable of breaking the hard Brazil nuts and dispersing the seeds. Thus, management of hunting may have positive results on both the fauna and the flora, assisting the maintenance of viable populations and ecological processes. These are necessary for the forests to continue to provide the necessary ecosystem services and provide local people with secondary income as well as a safety net in times of environmental, economic and social stress.

Management of Forests and Trees within Landscapes

Adaptation strategies geared towards communities will in general prioritize food and water security as well as health. In rural landscapes, forests and trees may have an important role in securing food, water and health. Land owners will benefit from mutual collaboration at this scale, identifying priority areas for the protection of water resources, strategies that build a common emergency reserve of seeds and food supply, measures that preserve or improve essential ecosystem services (such as pollination or water quality and soil protection), and measures that will reduce the breeding sites of potential disease vectors.

Although not necessarily with the explicit intention of adapting to climate change, several territories adhered to the International Model Forest Network that promotes the formulation and implementation of such strategies. In 2010, 58 model forests existed worldwide ; 11 of them were implementing specific climate change activities at a landscape scale, while another 9 were working on environmental services. In general, the model forests are seen as good opportunities to implement international agreements on the ground. Their main strength being that they foster local governance and collaboration through voluntary participation in discussion platforms of the local actors. Worth mentioning in this context is the Green Belt Movement, that has shown over the past 30 years that tree planting can contribute to rehabilitation of degraded lands, increased yields of small holder farmers in thousands of communities, while at the same time contributing to CO2 sequestration (Green Belt Movement).

ADEQUATE MANAGEMENT RESPONSES TO CLIMATE CHANGE

The enabling conditions for the forest management responses to climate change are identified and the challenges that these gaps may create analysed. It should be noted that relatively few experiences in forest management responses to climate change have been documented and therefore it is difficult to provide a comprehensive overview of existing capacities and those needed for adaptation and mitigation actions. Lack of knowledge on climate change impacts on forests In many tropical countries, information on forest resources (type, quality, extent, values, and changes) is deficient. This information can come from studies on specific topics, (*e.g.* tree growth of individual species under various conditions), or from local or national monitoring, gathering data with a predetermined frequency on a predetermined set of variables relevant for forest management and conservation. For the more general information, such as forest area, forest quality and species composition, monitoring is the preferred method. In Europe, the United States and some tropical countries, such as Mexico and Chile, such information comes from continuous national forest inventories. In most tropical countries, however, such monitoring mechanisms are limited and sometimes non-existent.

NATIONAL MONITORING SCHEMES

In a revision of national monitoring schemes related to regional SFM standards (such as the Tarapoto process, the Montreal process and the ITTO criteria and indicators for SFM) at a workshop in Chile in April 2011 (Günther *et al.*, in prep.), many similarities were identified between the standards at the criterion level, but countries found it difficult to harmonize monitoring at the indicator level. This was partially due to differences in priorities, but largely also due to differences in capacities to measure the indicators and in the institutional arrangements necessary to measure and share the results between the different state and private entities involved.

This is in particular the case for socio-economic data, but also for the quality of forest resources. The quality of monitoring, furthermore, depends on continuity of the measurements, which requires long term funding (often not available in tropical countries), clear and constant definitions of the variables and processes to be measured (such as forest degradation and deforestation and forest quality) as well as compatible measurements over time (also not available for many indicators). The FAO Forest Resource Assessment (FRA) process was recognized as providing a good base to set up and harmonize monitoring tools and mechanisms. However, it was noted that this process suffers from the same obstacles as the existing national and regional processes.

At the project or community scale, monitoring often encounters even greater obstacles, due to the disadvantages of measuring at smaller scales. While it may be possible to simplify monitoring mechanisms, this is only possible if the monitoring results are adequate for the scale, intensity and risk level of the proposed forest operations. Even in certified operations, where monitoring at project scale has developed further than in most other operations, only few well designed and operational monitoring systems exist that actually contribute to improved forest management.

While such monitoring is essential for good forest management at any scale, it needs to be accompanied by monitoring of weather conditions in order to be a useful tool for forest managers to prepare themselves for climate change. In many countries there is no detailed information on weather conditions at the local scale, limiting the ability to relate past and current forest behaviour to weather conditions. In addition, the lack of such data reduces the capability to downscale climate change projections and thus the ability to project climate change impacts at a local scale. It is therefore necessary to increase the number of weather stations throughout much of the tropics (especially the large forest areas), as well as the capacity to analyse the data and incorporate them into impact and climate change models at a local scale.

Research

Research is usually used to solve specific problems or establish links between actions and impacts. It may also be used for a more in-depth understanding of forest processes and the contribution and influence of human activities on forests. Research on climate change impacts and adaptation options have been focused mainly on boreal and temperate forests. Little information exists on impacts of climate change on tropical forests and how tropical forests will react. While potential changes can be studied in field and laboratory trials, it will also be necessary to improve climate change projections at a small scale. Only with reliable local climate change projections will forest managers be able to plan for future changes. Data on the response of species and individual trees to changing environmental conditions is particularly important for plantations, where a limited number of species is planted and thus substitution may not take place when conditions change. For the major timber species, more work is required to identify their levels of resilience to climate change. This information will also be relevant for mitigation, since tree dynamics and its change under changing climatic conditions will influence the capacity to sequester carbon. Research on functional groups may be as relevant specifically for adaptation of forests. There is a lack of case studies highlighting successful cases of forest managers implementing adaptation strategies. Currently, most studies look at tree or animal species. Few have a system approach that includes the environmental, economic as well as the social and technical aspects of forest management.

Much is already known about the relationship between trees and water. However, this relation may vary according to geographic, topographic, geological and soil conditions. More studies are required to understand local influences of trees and forests on the hydrological cycle, especially in those areas where climate change is likely to change the amount and the distribution of rainfall over time. Climate induced changes in the water cycle (*e.g.* increased frequency and severity of droughts and floods), may affect the capacity of local forests and people to adapt to climate change as well as the mitigation potential of forests. Further, there are several myths surrounding the beneficial effects of trees and the negative effects they may have on water availability, soil erosion and other potential ecosystem services. Undoing those myths may be harder than learning from new experiences.

The social and cultural impacts of climate change on forests dependent communities have not been studied widely. Changing forests may change the availability or quality of ecosystem services, resulting in changes in the relationship between people and forests, social relations among people and the degree to which cultural needs are met. This could lead to migration to other areas in search of the lost ecosystem services or other livelihood opportunities. One of the consequences of migration is increased conflicts over resources. There is an urgent need to study these potential conflicts and to recommend options to manage these conflicts. REDD+ is being widely studied, particularly at the project and national levels. The impact that REDD+ strategies may have on individual households in the medium and long term needs to be further studied to understand the potential contribution that REDD+ may have to local development and biodiversity conservation. While this may not be of direct concern to forest managers, it will, in the long term, be important to avoid social conflicts and the results of such studies may alter the REDD+ mechanisms that are in the process of being designed.

Communication

Monitoring and research will be of little use if the results are not communicated to potential users and the general public at large. IUFRO, among others, have published guidelines for effective communication in the forestry sector and there are many cases where these guidelines have been successfully applied. In general, however, communication within the sector is weak, resulting in many stakeholders not having access to reliable and updated information. They may miss new opportunities or even act upon incorrect information.

CAPACITIES OF FOREST MANAGERS TO RESPOND TO CLIMATE CHANGE

Many forest managers are aware of climate change and that it may have positive or negative effects on their forests, but only few have a good understanding of the degree of uncertainty of climate change and impact projections. They are thus susceptible to wrongly interpret those projections unless they receive capacity building on forests and climate change. Most of the existing and proposed actions for mitigation or adaptation to climate change do not require different technical capacities to those already required for SFM. However, it is often not clear to managers that existing best practices may be adequate for the expected or occurring climate induced changes in the natural environment. Even without considering climate change, many forest managers are not up to date with the best practices for SFM. Preparing for climate change, therefore, needs to include capacity building in SFM practices, including integrated fire management, integrated pest and disease management, harvest planning (including road building), silvicultural practices and the implications actions for nutrient and water cycles. The practices to be focused upon during capacity building will depend on the country and target group. In the Mediterranean, for example, training for fire management will be more relevant whereas in tropical American pine forest, pests and diseases may be a more relevant topic, although fire management is also important in these areas. In addition, there is a general need to strengthen the monitoring capacities of forest managers. Awareness raising on adequate monitoring methodologies, tools, and indicators and threshold values that will allow for early detection of changes is also required. This is relevant especially for managers focussing on timber production, forest conservation and protection. Although the livelihoods of forest dependent people are largely based on NWFPs, there is relatively little awareness of the ecological requirements and management practices of these NWFPs. Capturing that knowledge and sharing it with others will greatly contribute to strengthening adaptive capacity of many communities.

Appropriate Technology

Appropriate technology is needed for three major lines of action: monitoring and research, adaptation and mitigation and will vary according to geographic area, type and objective of forest management activity, scale and intensity of operation and existing local human and financial resources. The major problem with the availability of technology is not its existence, but its location and adaptiveness to local conditions. It may be as simple as having instructions for the use and maintenance of equipment only in a foreign language, not readily understood by the intended users. The cost for acquisition and implementation of this technology is another common problem.

Monitoring and Research

Monitoring and research of vegetation cover usually requires a range of instruments, from simple measuring tapes to highly sophisticated remote sensors. In addition it requires specific software that allows storage and analysis of the data. While instruments and software exist, they are often not accessible for, or need to be adjusted to the requirements of forest managers. The experience with software for forest inventories shows that tailor-made software has a tendency to get out of date rapidly and after a few years can no longer be used within the new operational systems. It is therefore becoming more common to look for commercial software. The large Canadian based forest products company Tembec, for example, has used a package of commercially available software as basis for their ISO 14001 monitoring system. This, however, requires qualified personnel to combine the software and adjust it to the company's needs. Small and medium scale forest managers, especially in developing countries, will need to collaborate with research and monitoring organizations to be able to make the necessary adjustments in a cost effective way. Adaptation may require new irrigation technology, as well as breeding technology and seed storage technology. Mitigation, on the other hand, will need technology efficient in energy use.

GAPS IN THE INSTITUTIONAL ENVIRONMENT

For the purpose of this document, the institutional environment is considered the framework of formal and informal arrangements that will set the limits for the implementation of climate change preparedness activities.

This environment includes the normative framework (legislation and policies), markets and other arrangements to finance forest management and conservation, as well as informal arrangements that direct behaviour of the local actors.

Property Rights

Ownership and property rights are a major concern in many regions of the world, particularly for indigenous people in developing countries. There are some cases where rights and tenure are protected by legislation but enforcement is oftentimes lacking. In Costa Rica, for example, land, tree and carbon rights are well-defined, but still a considerable part of the defined rights in the northern zone of Costa Rica are currently under dispute. In other countries, poor people are characterized by not having legal access to land.

Normative Framework

While many countries have recently revised or developed forest legislation, the normative framework for forests and climate change needs to go well beyond legislation and policies of the forest sector. Commerce, transport, agriculture and finance are just a few of the other sectors that have a great influence on what happens in forestry. Mechanisms to foster this intersectoral collaboration are often nonexistent in most countries and often climate change is dealt with outside of the forestry sector.

The different actors of the forestry sector are often not well-organized at the local, national or international levels and have little experience in transparent policy setting and implementation. In many cases this has led to a lack of trust, which makes collaboration even more difficult.

Financial Arrangements

Investment costs of change are high, while at the same time forest owners or users do not have access to financial mechanisms that help cover those costs. REDD+ talks have created great expectations amongst developing countries however there have been many obstacles thus far for the successful implementation of REDD+ activities. It is thought that the costs for REDD+ implementation may exceed the capacity of developed countries. Further, developing countries do have the absorption capacity for the REDD+ funds. This may leading to uncontrolled spending and may do more harm than good in the long term. Markets for carbon from trees and forests are still not well established and there are no generally acceptable and workable definitions of permanence, additionality and leakage or measuring and monitoring systems in place. Few effective PES schemes exist, mainly because the institutional structure is not in place or too expensive to operate.

8

Carbon Crediting and Forest Management

INTRODUCTION

Deforestation in the tropics is known to be a major source of carbon emissions and an active contributor to global warming. The IPCC estimates that 1.7 billion tons of carbon are released annually due to land use change, of which the major part is ascribed to tropical deforestation. This represents 20-25% of current global carbon emissions. Deforestation emissions from Brazil and Indonesia alone are equivalent to the entire reduction commitment of the countries during the first commitment period. *Degradation*, the loss of biomass from within the forest as a result of thinning out of the vegetation, is also a major source of carbon emissions, but statistics on its incidence and on the associated carbon losses are virtually non-existent. Under the current agreements in the Kyoto Protocol and the Marrakech Accords, neither deforestation nor degradation of tropical forest are addressed. Possibilities under the CDM are limited to afforestation and reforestation, and do not include management of natural forest. In other words, they allow for planting of new trees to establish additional sinks, but they do not allow crediting for reduction of emission from existing sinks.

In response to calls from a number of Parties, the UNFCCC at CoP11 in December 2005 initiated a two year process for the consideration of a policy for "reduced emissions from deforestation". This debate is on-going, and covers political issues, methodological challenges, such as how to measure and include degradation, and alternative financial mechanisms that might be employed if such a policy were to be adopted. The Kyoto: Think Global, Act Local project has been working since 2003 to develop methods and to make policy suggestions in this area. This booklet explains the rationale and presents preliminary findings on the basis of six case studies from sites in Africa and Asia.

CARBON STOCK CHANGES AS A RESULT OF MANAGEMENT ACTIVITIES

As a result of participation in the Kyoto: Think Global Act Local project, five members of the Forest Committee (three men and two women) were trained in mapping techniques using GIS/GPS on a hand held computer and in standard forest inventory methods as described in the IPGG Good Practice Guide. They established 19 sample plots of 5.6m radius, laid out at intervals of 218 meters using transects separated by 286 meters. Locally derived allometric equations were used to calculate the total biomass and to convert this into tons of carbon stock. Below ground carbon stocks were not estimated but in principle could be calculated and added to the total.

The stand parameters for Handei village forest reserve. Observed stem numbers in this forest are comparable to other forests in similar (protected) site conditions while volume, biomass and carbon per hectare are generally lower. This is probably because the forest is still regenerating following previous disturbances including agricultural fields with few trees. However, analysis of data between 2005 and 2006 shows that the forest is growing and has sequestered about 3 tons of carbon per hectare in the year interval between the two measurements. Data for several more years will need to be collected before a growth curve can be drawn, but the evidence is clear: the forest is increasing in carbon stock as a result of the management practices used by the villagers.

The managed forest clearly shows an increase in carbon stocks due to the suppression of unsustainable harvesting of fuelwood and charcoal, of around 5 tons CO_2 per hectare per year. The village forest management regime is thus sequestering a considerable amount of carbon as shown above. From the data so far available, it is not clear to what extent emissions are being reduced in addition, since the rate of depletion of forest in the unmanaged area has not yet been established. In order to make an accurate assessment of this, data over several years will be required, and any leakage from the managed area will have to be accounted for. It is the intention of this research project to continue monitoring carbon stock changes to establish annual rate of carbon loss and predict future carbon stocks. This will form the baseline scenario against which carbon benefits of the reserved forest will be compared.

Deforestation and Degradation

There is no doubt that deforestation, the full conversion of forest land to other uses, is occurring on a large scale in many countries and images of this – fire-devastated hill slopes, massive chain saws felling large buttressed tree trunks in tropical jungles – appear frequently in the popular media in an appeal to people's innate love of nature. To counter deforestation effectively however it is important to understand the underlying causes and drivers. Much deforestation is the result of planned activities which are necessary for development. It is an inevitable (though regrettable) side effect of rational choices that are made by governments and individuals, which bring about land use change for the sake of greater production. The expansion of area under cultivation for food crops and under pasture may be a priority for economic growth, for feeding the growing population and for earning export income. Conversion of forest to plantation crops increases national income. Logging provides essential funds for investment in development. Cities grow and infrastructure is constructed as part and parcel of modernization and the increasing scale of the economy. These are *governed* activities, which for the most part cannot and should not be stopped; they are essential for development. At best, the impact on forests could be softened by ensuring good coordination between sectors and overall land use planning, the use of more sustainable timber extraction methods, and the encouragement of agricultural systems which retain as much carbon as possible.

However, there is a great deal of what might be called *'ungoverned'* deforestation going on as well. This is deforestation which is not sanctioned, and usually takes place at the frontiers of the forest. The stakeholders are individual farmers or small agricultural concerns working more or less on their own accord, although in many cases an 'agent' organizes the deal, and it sometimes occurs with corrupt complicity and a 'blind-eye' from local authorities. It mostly involves agriculture but in some places illegal logging is the main cause. Many countries find it very difficult to control this kind of deforestation, which is driven by market incentives and lack of alternative opportunities, and thrives on weak enforcement of law and lack of government capacity.

STOCKS OF BIOMASS IN THE NATURAL FOREST

Degradation - the gradual reduction of stocks of biomass within the natural forest – is however a quite different process. Degradation results from extracting more biomass from the forest than it can sustainably produce. Levels of biomass – and therefore of carbon – dwindle; slowly at first, but gradually the forest thins out more and more until one could say that the area is really deforested. Often this is not the result of a single or coordinated and rational decision to clear the forest, but of a number of processes that have to do with the livelihoods of people nearby. Grazing of cattle within the forest prevents regeneration of saplings and shrubs; over-harvesting of wood for the production of charcoal to sell in the cities overstresses the productive capacity of forest; slash and burn agriculture, a traditional and normally sustainable forest land use, becomes devastating if the fallow cycle is too short to allow the forest to recover. Local people are well aware of the impact of these activities on the forest and of their negative implications. There are two sets of reasons why they continue to carry them out. Firstly, there is usually no alternative means of making an income, and secondly, the forest is to all intents and purposes an uncontrolled resource. The majority of the forest is owned by the state, but apart from heavily protected areas such a nature reserves, most is *de facto* open access. With no rules for usage, or no enforcement of rules, each individual makes the most of his or her opportunity, because if not, someone else will – *the tragedy of the commons* – or, as it may more correctly be described, the *tragedy of the open access resources.*

GLOBAL LOSS OF FOREST BIOMASS

How much of the global loss of forest biomass is due to full deforestation and how much due to creeping degradation? This is difficult to know, not least because most countries do not monitor degradation at all – it is not easily visible from remote sensing – and therefore do not report it to FAO. Tropical rainforest has very high carbon densities (up to 400 tons per hectare) but forms only a small proportion of all forest area. It is threatened by large scale deforestation in some areas, most famously in the Amazon. The vast majority of tropical forest is dry, with carbon densities of 40-80 tons per hectare. Some of this, particularly around cities, is being cleared wholesale, but much of the rest is subject rather to degradation. The processes are not entirely independent, but they tend to be focused on different sorts of forest and in different geographical situations. What we can say is that both deforestation and degradation contribute significantly to global carbon emissions, and for that reason *reducing emissions from deforestation and degradation (REDD)* is the most appropriate general term for actions designed to curb these processes.

COMMUNITY FOREST MANAGEMENT FOR REDUCING DEGRADATION

In recent years the inability of the state to control degradation of forest has been recognized in many countries. Governments are seeing the benefits of handing over forest areas to local communities under a variety of community forest management schemes, in India, Nepal, Papua New Guinea, Burkina Faso, Tanzania, Cameroon, Mexico, Peru and many other countries – it is estimated that around 14% of all forest in developing countries is under this kind of management today, three time more than 12 years ago. Under such schemes, villagers get the formal, legal rights to use and profit from the forest products, under jointly agreed management plans which ensure that off-take is kept at sustainable levels. Communities organize themselves by setting by-laws and by self-regulation as regards access to forest products. Their motivations to take part in such a scheme can be various: to maintain the forest to ensure future benefits is a clear overall reason. For some, it is to ensure a continued supply of firewood and fodder; for others, to enable eco-tourism; yet others participate in the hope that the wild animals that have disappeared from the shrinking habitat will return, and provide a means of sustainable subsistence in the future. In a few, sustainable timber off-take is the aim. The benefits are usually small in financial terms, but real and tangible in non-monetary ways.

Initial experiences of such community forestry are by and large positive. Areas which are community managed are clearly distinguishable from surrounding areas which are not; as natural regeneration appears to be taking place and biomass is more dense, so that instead of being a net emitter of carbon, the forest becomes a sink. Furthermore, it is probable that without such management, the biomass would decrease, through forest degradation, leading to additional carbon emissions. As the case studies in this book show, the gains could be anything from 4 to about 12 tons of CO_2 per hectare per year, depending on the type of forest.

CARBON SERVICES ACT AS A STRONG INCENTIVE AGAINST DEGRADATION

If carbon has a monetary value, could payment for reduced emissions from deforestation act as an incentive for this kind of forest management activity at the local level? Would it stimulate more communities to adopt simple management rules over much larger areas of natural forest, to bring rates of extraction into balance with the forests' natural capacity to reproduce? If this were the case, then many parts of the forest in tropical areas might be involved in reducing carbon emissions, and very many small communities might earn some income from this new service. Naturally, there would be many additional positive side effects, not least the maintaining of biodiversity, water management, erosion control and the fight against desertification. It is clear that the attractiveness of this kind of option to local people will depend greatly on the opportunity costs of keeping the forest as forest. In areas where an alternative land use – plantation, or pasture – is likely to give high financial returns, then it will be difficult for carbon to compete; these areas are likely to be deforested, come what may. But in more remote areas, particularly drier areas where agricultural production potential is low, there could be a real niche for 'community carbon forestry' targeted at reducing and reversing degradation.

In order to assess this possibility in more depth, it makes sense to look carefully at community forest management experience and evaluate its impact on carbon stocks. There a number of questions that would need to be addressed, such as:

- What rates of degradation and carbon loss are typically occurring in unmanaged forests?
- What sorts of management activities are used by communities under CFM schemes and how much carbon is saved as a result?
- Is there leakage to other areas? How much?
- What is the opportunity cost of this management?

- How could the carbon stock changes be measured and monitored in a cost-effective manner?

The 'Kyoto: Think Global Act Local' research project, funded by Netherlands Development Cooperation, has set out to answer these questions and to assess the potential for community carbon forestry. Working with local NGOs and research institutes in Mali, Senegal, Guinea Bissau, Tanzania, Uganda, Nepal and Uttranchal (India), communities already engaged in local forest management have been trained in the use of a small handheld computer with GPS and GIS equipment which enable them accurately to map the boundaries and the strata in the forest – a prerequisite if the carbon savings are to be verifiable. Further they have been trained in standard forest inventory methods, using fixed sample plots, and in entering this data into a tailor-made database on the computer. None of these villagers has more than 7 years of primary education, and none of them has ever seen a computer before, but this is no hindrance. The local NGOs help in the training, maintain the computers and supervise the laying out of the sample plots to ensure that the carbon measurements meet rigorous scientific standards.

If such an approach were to be considered viable, then a number of further questions need to be posed:

- When local people measure and monitor carbon stock changes, are the results reliable? What technical problems arise?
- What is the cost of such an exercise (local transaction costs) in relation to the amount of carbon generated? Is the exercise worthwhile in the eyes of the local people; what level of payment would be necessary to make it worthwhile?
- What will be the impact of carbon payments on other forest values, and on the social network? Who will benefit, who will lose?

VILLAGE FOREST RESERVE

Community Forests Management initiatives were introduced in Tanzania in the early 1980's with some experiences of success stories from Nepal and India. The practice is already legitimized by the parliament through the current forest act (2002). Under this act there are mainly two main ways in which communities are involved in forest management: these are Joint Forest Management (JFM) and Community Based Forest Management (CBFM). Under JFM, the government involves local communities in carrying out different forest activities (such as patrolling, fire fighting and boundary clearing), as such forest ownership remains with the government while local communities are duty bearers and in turn get use-rights and access to some forest products and services. On the other hand in CBFM the local communities are the owners, as well as right holders and duty bearers. Most of the CBFM forests are demarcated as part of village general land. Thus they are also called village forest reserves. To date there is a total of 994 different areas involving 2009 villages with a total area of about 3 million ha under community forest management in the country. However, current statistics also reveal that the remaining forest area in general land is about 18 million ha. These forests are "open access" characterized with insecure land tenure, shifting cultivation, harvesting for wood fuel, poles and timber, and heavy pressure for conversion to other competing land uses, such as agriculture, livestock grazing, settlements, industrial development. In addition, the lands are subject to wildfires which are caused by human activity. The rate of deforestation in Tanzania which is estimated at more than 500,000 hectares per annum is mostly impacting such general land forests. Therefore there is a room for many more community forest management activities that may alter the observed high rate of deforestation in the country.

HANDEI VILLAGE FOREST RESERVE

Handei village forest reserve is located in the Eastern Usambara mountains in Tanga region and is just outside the Amani Nature Reserve. It consists of 156 hectares of sub-montane evergreen forest characterized primarily by *Parinari excelsa, Sapium elleplicum, Cynometra sp* and *Alanblankia stulhamanii* species. Part of the forest is on hanging rocky cliffs harboring *Saintpaulia usambarensis* (African Violet) species that attracts ecotourism. The forest has been under community based forest management by residents of Magambo-Miembeni village since 1996. Formerly, the forest was under open access and suffered considerably from agricultural expansion and uncontrolled harvesting mainly for commercial timber and building material, the consequence of which were changes in microclimate of the area and drying up of important water sources to the local communities. With current management, utilization is confined to a buffer zone of 50 m from all sides of the forest boundary, the interior part of the forest is for protection without utilization. Uses permitted in the buffer zone include: ecotourism, timber harvesting, collecting dry firewood, vegetable, mushroom and collection of traditional medicines. To ensure proper utilization, the village has set down various bylaws on how and when these forest products can be utilized, the general idea being that utilization is done in a sustainable manner.

There is a village forest committee composed of twelve members (currently 4 women and 8 men) operating under the village government that manages the forest. The committee is responsible for all activities regarding the forest, these include: selecting forest guards, monitoring of all activities conducted in the forest such as enrichment planting in open areas of the forest, provision of permits for various activities such as harvesting of timber and collection of fees from ecotourism. It is also responsible for following up on legal issues pertaining to the management of the village forest reserve. The committee reports on a monthly basis to the village government, district forest officer and a local supporting organization (the Amani Nature Reserve conservation office). The role of the district forest officer and the supporting organization is to provide technical support to the forest committee and interpretation of policy guidance.

FOREST AREA

Kitulangalo forest area lies about 50 km to the east of Morogoro town, on the side of the Dar es Salaam-Morogoro highway. This is a relatively dry area with an average annual rainfall of about 850 mm. Formerly the forest was part of the Kitulangalo Catchment Forest Reserve. The high level of accessibility to the highway made this area a prime charcoal production area for the supply of the nearby Morogoro municipality and Dar es Salaam city. But in addition the forest suffered from timber extraction through the activities of local pit-sawyers, and from cutting of tree stems for building poles. The human resources of the Forest Department were insufficient to maintain control over the area and to prevent the over use of this important catchment forest. It was *de facto* an open access resource.

In 1995 however, part of the forest (600 ha) was made over to Sokoine University of Agriculture (SUA) as a Training Forest Reserve; it is now used for training students and for research purposes, although protection was a major reason for its new status. This part of the forest is under joint forest management with Gwata village, which means that the land is still owned by the government, but the management is mainly in the hands of the local community, following jointly prepared management guidelines. In 2000, another 420 ha was demarcated for the village community, and is now called Kiminyu village forest reserve. As a community forest, the land is now the property of the village, which has full responsibility for management. Both areas are characterized by Miombo (savanna woodland) and the predominant species are *Brachystegia* and *Julbernadia.*

DIFFERENT MANAGEMENT STRATEGIES AND RULES

The fact that two different management regimes are operating next door to each other in essentially the same type of forest makes the Kitulangalo forest a particularly interesting one to study. In Gwata village an environmental committee has been established and given the responsibility for supervising the management of the forests on behalf the village government. This committee has been established to look after all forest management activities in the villages. The committee members are selected by village government and approved by general village assembly. The considerations for selection to the committee are; village residence, married person, and ability to work. Intrinsically, gender balance is also carefully considered in order to involve women in the management of the village forest reserves. To institute its mandate, the committee sets up bylaws that are approved by the village general assembly. These bylaws are also approved by the responsible district authority and are recognised by the court of law. They consist of different penalties charged against offenders who violate the rules regulating sustainable forest management and use in the village. These bylaws are applicable in both village and government owned forests in the village.

Sokoine University manages the training forest jointly with the village government through the village environmental committee. Two members from the committee are employed by the university as forest guards for the forest. These are responsible for making routine patrols and they supervise different silvicultural activities that are done by villagers who receive daily wages in return. For example, the university involves villagers in clearing of forest boundaries to safeguard against fire. This is normally done during the dry season when the grasses are dry and vulnerable to fires. In the same boundary lines, villagers plant trees, which are used to demarcate the reserves and general village land. If there is fire outbreak, the villagers are also involved in extinguishing it. However, it may be noted that incidences of fire outbreak in the Training Forest Reserve have considerably reduced in recent years since local people have been involved in forest management.

The village environmental committee bears full responsibility for managing the village forest (Kimunyu Forest). It mobilises local people, and selects villagers to patrol the forests every day and report to the village government through the committee. Although this forest is being managed for production purposes, currently there is no tree harvesting allowed. There are not yet enough large timber tree species in the forest, and the only product that could be extracted at present would be charcoal. However a decision has been made to stop charcoal production and to allow the forest to regenerate naturally. The result is that currently, there are higher trees stocking levels in this forest compared to the adjacent public land that is under open access management.

GROWING CARBON STOCK

The improving health of the forest can also be seen from the point of view of carbon stock. In Gwata village, 6 persons (4 women and 2 men) were trained in mapping and forest inventory techniques as in all the other study sites under the Kyoto: Think Global Act Local project, with the help of two forest guards who are employed in connection with training forest reserve. In the Training Forest Reserve, 89 plots were set out at intervals of 150 meters along transects set 300 meters apart: in the Kiminyu village forest reserve, 43 plots were set out at distances of 170 meters, on transects separated by 500 meter. The number of sample plots was in each case calculated based on estimates of standard error, based on preliminary sampling as outlined in the IPCC Good Practice Guide and the Winrock/Biocarbon Fund Sourcebook.

To draw firm conclusions concerning rate of carbon sequestration, data over more years will be required. However it may be borne in mind that, had the forest been left without community management, carbon stock would certainly have decreased, as had been the pattern over earlier years. The rate of forest loss and of degradation can be determined from studies that were carried out in areas in the vicinity of Kitulangalo, which show that the rate of loss of forest is strongly related to distance from the highway. Over a period of 6 years stock levels dropped by as much as 80% in sites up to 5 km from the highway, but only by 20% at 10 km. This is the result firstly of charcoal production and later of wholesale clearance for agriculture.

The increase in standing volume at 15 km is due to the fact that this area is now under community management for some years (this is the area that is now Kimunyu village forest).

If a conservative estimate of 5% biomass loss per year was to be assumed as the average baseline, then the net gain in carbon terms as a result of community forest management would be on the order of 10 tons per hectare per year. At a nominal value of $2 per ton after deduction of external transaction costs (ie non-local costs involved in verifying and certifying the carbon gains), this would be equivalent to an annual income of $20 per hectare or $8400 for the Kimunyu forest alone. It might be expected however that there is some leakage, in the form of displaced activities, from these sites. Villagers in this village collect firewood and building materials from the general land that is at close by distances from their homes. Only tree felling for commercial timber extraction and for charcoal making could be assumed to be displaced somewhere else. However, there are no evidence of villagers' migration to other areas to deforest. Of course, it could be argued that the charcoal may still be produced elsewhere, by other people, to meet the urban market demand for this vital product, and thus represents a form of leakage, but it is difficult to prove this or to estimate its impact.

Local Transaction Costs

Measuring biomass stock to determine changing carbon levels itself involves costs, which are considered to be local transaction costs. At Kitulangalo the costs involved were recorded. A comparison of costs of carbon assessment by local communities against the professionals reveals that it costs twice as much to hire professionals for carbon assessment in the village forests studied, as to engage villagers to do this, including the cost of technical assistance and training, which is considerable in the first year of assessment. It is to be expected that the villagers will be able to undertake the same work at progressively lower cost in the preceding years as the cost for training and supervision are reduced. It is assumed that from the fourth year, the villagers can work on their own with assistance only from staff from their local supporting organization. It is also clear that it is more cost effective to work with villages which are managing large forest areas, since the cost of training is a fixed cost.

PROMINENT ROLE OF COMMUNITY FOREST

Community forest plays a prominent role in the hills of Nepal where agriculture and livestock rearing and forest are strongly interlinked. Based on the 1976 National Forestry Plan, the government of Nepal made a policy to involve local communities in forest management, with a view to tackling deforestation and the deteriorating state of the forest all over the country. By 2004 about 25% of all national forests, or around 1.1 m ha., were being managed by Community Forestry User Groups (CFUGs). There are more than 13,000 CFUGs in the country, involving 1.4 million households (*i.e.* 35% of population) , mostly in the hilly regions of Nepal. The Federation of Community Forest Users Nepal (FECOFUN) has grown over the years to become the largest organization in the country. The impact of this policy in the forestry sector has been positive. Where communities are managing their forests, the degradation trend in the hills has been checked. Forest conditions have improved in most places with positive impacts on biodiversity conservation. Communities have easier access to firewood, timber, fodder, forest litter and grass. Soil erosion has been mitigated and water sources have been conserved in such areas.

As a general rule, members of the CFUGs pay a nominal fee for the various forest products they consume and are restricted from harvesting of forest products for commercial purposes. Timber harvesting in particular is heavily regulated and only conducted under Forest User Committee (FUC) supervision; selling is done through an open bidding process. All income from such sales is retained by the CFUG. Revenues collected by the CFUG from the members and through selling products are mostly reinvested in social infrastructure as requested by the community members. About 28% of the revenue generated from the community forest is expended on forest protection and management.

This case study looks at one example, the community forest in Lamatar, to demonstrate that in addition to other forest benefits, community forest management results in increasing carbon sequestration and also quite probably in decreasing emissions.

HISTORY OF THE KAFLEY FOREST

Lalitpur district has 15,253 ha of forest of which 9,993 ha are managed by 162 CFUGs. Kafley Community Forest is one of these. It is a block of 96 ha which is being managed by the Kafley CFUG, which consists of 60 households. This forest lies at an elevation of between 1,830 and 1,930 meters and is dominated by temperate broad-leaved species, particularly *Schima-Castanopsis* (katus-chilaune). The tradition of community managed forest here is not new, what is new is the formalization of the traditional management practice in modern terms.

Villagers recalling the history of their forest management explain that the forest in the Kafley area historically belonged to the Ghimere family, who were Brahmins living to the south of the main valley. They had agricultural lands in the fertile valley below the hills; the hills themselves were unsuitable for agriculture and were covered with forest.

They were granted this forest as *Birta* by the State for services rendered. It is told that the forest was rich in biodiversity at that time, as it was well managed. In 1957, however, this forest, like all forests in Nepal, was nationalized. After that, as narrated by the locals, the forest gradually decreased, both by outright deforestation (loss of forest area) and in terms of degradation (loss of biomass within the forest). Noticing this change, the Department of Forestry carried out a reforestation programme in 1978 by developing a sallo plantation (*Pinus roxburghii*) and putting forest guards in place to protect it. But deforestation and forest degradation continued unabated, converting the entire hilly area to almost barren land by the early 1980's. Unregulated livestock grazing and fodder collection were the major causes of forest degradation as they prevented natural regeneration, while unrestricted fuelwood and timber collection were the major cause of deforestation. This was a classic case of the *tragedy of the open access*; anyone and everyone had unlimited access any time because the state owned the resource and it was managed by their staff, to whom the local people did not feel answerable.

The scenario at Kafley was occurring all over the country which meant that Nepal was losing forests at a rapid rate especially in areas adjacent to settlements. In the late 1970's however a paradigm shift occurred, when foresters began to realize that forest protection and management was not possible without involvement of the local people. Between 1975 and 1993, a series of milestone decisions brought about the community forestry policy that we see practiced so widely in Nepal today. Most of the handing over of forests to the local communities took place in the 1990s. In Lamatar this happened in 1994, a year after the formation of the Kafley Community Forest User Group. Since then, forest has been managed effectively with strict restrictions and user guidelines and norms. Forest degradation and deforestation have been checked and forest regeneration (which is mainly natural regeneration) is taking place after stringent protective measures were deployed by the local people through the CFUG. Today the forest is recuperating ecologically and already has a rich diversity in tree species. One of the most important resources obtained from this forest is water. This forest has several springs which are carefully protected and used by the village for drinking purposes, at no charge to the users. It has been reported that the flow of water has markedly increased with the rejuvenating forest ecosystem.

Management Regime

Membership of the CFUG is not compulsory but all villagers who need forest products are members, to ensure their access to the forest. The Kafley CFUG has a constitution and a five-year operation plan that indicates how and for what purpose the forest will be managed. The CFUG is headed by a Forest User Committee (FUC) consisting of 11 elected executive committee members (of whom 6 women), which makes day to day decisions and calls the CFUG meetings. The primary mission of the Kafley CFUG is to increase the harvesting capacity of fuelwood, timber and fodder through better management of forest resources for the benefit of the local CFUG members and to make the CFUG a self-sustaining institution. But in addition, the CFUG aims to conserve spring water sources, soil and biodiversity and promote environmental stability in their village area. The CFUG also assists in raising living conditions from the use and access of forest resources, and is trying to develop this area for recreation and tourism uses. Community management of forest entails numerous tasks which the locals perform. Technical ones are undertaken with the support from the government forest rangers. Community management practices witnessed in the Lamatar area can broadly be classified into protection, administration, harvesting and forest management.

Protection is a major task and often the most expensive as well. CFUG has not hired anyone for patrolling the forest but is divided into subgroups taking the responsibility for patrolling on a rotational basis. While working at home or in the field below the forested hill, people keep an eye on the hillside and watch their forest for irregular movements, such as illegal logging, animal grazing or forest fire. In the past, people have been able to fight forest fires after seeing them from the field and rushing to the site immediately. It is compulsory for all members of the CFUG to participate in putting out fires, with penalties for failure in this regard. Penalties are in fact used for deterring all kinds of unsustainable forest resource extraction. Monetary fines are fixed by the CFUG meeting, with different rates for the illegal collection of fodder and litter, sand, gravel and stones, timber and fuelwood and bamboo, at times when such activities are not permitted. Hunting is permanently banned; grazing livestock and charcoal making likewise. Fencing as a protective measure is however not found here. It is the promulgation of these restrictions on use that has been the main management intervention and which has resulted in avoided forest degradation and deforestation.

The willingness of the community to implement these forest protection measures is related to and dependent on the pay-back they derive. It is clear to people in the Lamatar area that strict conservation measures, which are designed to maximize natural regeneration, in practice result in the harvesting of greater quantities of forest resources, and this is the incentive to cooperate in forest management under the CFUG. Community forestry also entails numerous administrative tasks such as calling and organizing meetings, conducting elections, recording and minuting meetings, maintaining accounts, getting accounts audited, etc, as well as those directly connected with forest activities such as setting dates for extracting resources and circulating the information, and developing the management plan and five-year operational plan with the assistance of a ranger. In Lamatar, such official administrative processes were found to be conducted rather professionally although not all CFUGs in Nepal are able to maintain such high standards in this regard.

Harvesting is done by all members. The main products extracted are timber, fuelwood (dried and green), fodder, litter, nigalo, (small bamboos: *Drepanostachyum intermedium, Drepanostachyum falcatum,* and *Sinarundinaria falcata)* and other non-timber forest products (NTFP). Of these, timber is the most heavily regulated; a decision to harvest is taken by the FUC together with the local forest range officer via an official process, and the timber is sold through a bidding process to anyone, including people from outside the village. Fuelwood, fodder, litter, nigalo and NTFP on the other hand can be collected by CFUG members when the forest opens; the FUC decides on the days and dates on which harvesting of these products is allowed in the different seasons and accordingly informs all CFUG members.

Members pay a small fee for firewood and bamboo, but fodder and litter are free. From records held by the CFUG, it appears that each household extracts about 1000 kg of green fuelwood, 500 kg of dry fuelwood, 500 kg of grass fodder, 1000 kg of leaf litter and 500 kg of nigalo every year.

On special occasions such a marriage, religious ceremony or funeral, 350 kg of fuelwood can be harvested by any CFUG member for the same price. Products extracted collectively after an operation such as thinning or clear cutting are distributed equally among the users. Members of the CFUG may sell any of their personal excess of these products to non-members within the village, but they may not be sold commercially outside the village. Sale of timber is the largest source of income for CFUG, followed by fuelwood fees. But unlike timber, fuelwood is extracted by the CFUG members only for fulfilling their subsistence needs and that of their fellow villagers, and though financially it is lower in value in terms of its contribution to the CFUG income, volume-wise it is the main resource extracted.

Most locals in Lamatar have their own clear understanding of silviculture as they have been interacting with forest even before going to school. Some of the locals can identify all the tree species in their forests, though the older men seem to be more knowledgeable on this than younger ones.

Some of the activities they conduct on a regular basis include weeding, cleaning, pruning/branch cutting, singling, thinning, clear cutting and regeneration management. The CFUG has maintained demonstration plots using modern techniques to propagate a number of species such as Chilaune *(Schima wallichii)* and Jhingane *(Eurya acuminate)* as well as several additional varieties of NTFPs (*e.g.* cardamom, fodder grass). In future Kafley CFUG intends to develop a forest nursery and also increase the number of medicinal plants in the forest.

Forest Inventory

As a result of participation in the *Kyoto, Think Global Act Local* project, members of the CFUG were trained in forestry inventory and mapping and conducted their own forest carbon stock assessment. Data from this is now available for two consecutive years. Table show very high number of stands and yet a low biomass per hectare (91.76 tha^{-1}) indicating that the forest is mostly at a young stage with vigorously regenerating saplings. However, in addition to the above-ground biomass as measured by the community, it would be possible to calculate the below-ground biomass using standard biometric equations, which would augment the annual carbon gains.

THE ORGANISATION OF FOREST USER GROUPS

A characteristic of the organisation of community forestry in Nepal is that the FUGs are socially heterogeneous, with members from both the dominant and the weaker social groups. The statutes require democratic decision making within the FUG, so this would seem to offer a vehicle for more participation of women and of poorer and marginalised groups and thus also an equal share in the benefits. The question is, whether this is the case in practice.

Several authors have suggested that women are not equally represented in FUG decision making, since each household is normally required to send one member to meetings, which in most cases will be the male head of household. Others (for example Nightingale, 2002) say that despite the principle of heterogeneity of FUGs, there remain power relations which result in more benefits reaching the more powerful members. In order to investigate whether these claims are valid, a case study was made in Baghmara Buffer Zone Community Forest in Chitwan, which is around 185 km to the south-west of Kathmandu.

THE COMMUNITY FOREST IN CHITWAN

Baghmara Buffer Zone Community Forest is in Bachhauli Village Development Committee (VDC), located on the northeast boundary of the Royal Chitwan National Park. The area is surrounded by the Rapti River in the south, the Budi Rapti River and Khagedi River in the northwest and the human settlements in the east. It is under the jurisdiction of Department of National Park and Wildlife Conservation (DNPWC). Prior to the handover of the Baghmara Buffer Forest as community forest it was heavily degraded and deforested by illegal activities such as timber felling, unsustainable collection of fodder, over grazing etc. Since this area was an extension habitat for the wildlife and in order to stop further degradation and deforestation and to conserve the forest, a plantation programme started in 1989 and in 1995 the DNPWC handed over Baghmara Buffer Zone Forest as a community forest to the people living near the forest area. Baghmara Buffer Zone Community Forest (BZCF) has 215 hectares comprising mono plantation, mixed plantation, natural regeneration, indigenous tree species such as sissoo (*Dalbergia sissoo*) and khayar (*Acacia catechu*), grasslands and lakes. The Forest Users Group (FUG) currently has 780 households as members, and these come from all castes and tribes: high caste Brahmins; middle caste Giri and Shresthas; low caste Darai, Pariyar, and Kumal together with people from ethnic groups or tribes (Bote, Majhi, Tharu, Tamang, Musahar and Magar). The Bote, Majhi and Musahar are the lowest in this social hierarchy; they are all well below the poverty line and are illiterate. For the members who joined at the start (in 1996) there was no charge for membership, but for new members, the membership fee is 3,000 rupees (wealthy class), 1,500 rupees (middle class) and 300 rupees (poor class).

INVOLVEMENT OF 'WEAKER GROUPS' IN FUG DECISION MAKING

Baghmara BZCF operates in accordance to its constitution and annual work plan approved by DNPWC. An executive committee is the apex body and is accountable for every activity that the FUG undertakes. Currently there are 13 members in the executive committee and these committee members were selected by the FUG members. The executive committee of Baghmara BZCF is socially heterogeneous and has representation from wealthy, middle and marginalised groups. According to the constitution of Baghmara BZCF, it is also mandatory to have at least two women members in the executive committee. Decisions made by the committee are first put in the general meeting and if two thirds of members agree, they are implemented. It is important to understand that in addition to daily management of the forest, the FUG is also responsible for the distribution of the forest products including any financial benefits that result from sale of forest products. In theory the executive committee works democratically and in a participatory manner, listing all the decisions to be made on an agenda for the general FUG meeting and accepting only those decisions that receive majority consent.

However, people of the Musahar tribe, a poor, marginalised group who are mainly involved in fishing activities, expressed their unhappiness as regards the composition of the executive committee. No Musahar has ever sat in the executive committee since the establishment of Baghmara Buffer Zone Community Forest. Currently, there are 23 Musahar households in the village and all live together in one part of the village in houses constructed by a Dutch NGO. Their children's education is funded by the same Dutch organisation. The adults in this group are illiterate and it is said to be for this reason that they have been excluded from the committee. They themselves do not often attend the general meetings of the FUG: they say that even when they are present, nobody listens to what they have to say. Their perception of the way the FUG works is that it is only nominally participatory, and that most decisions are made by the committee members or by the affluent members, and the general meeting is simply told what has been decided, rather than consulted.

There are 4 women on the executive committee, and these members are not from the high castes but from the better-off families of the marginalised groups. However, most of the decisions are made by the men members. The women have portfolios for particular tasks such as maintaining ledgers and organising meetings, and are involved in suggesting income generation activities that could be set up for other marginalised and poor women members, but weighing of the firewood during harvest and collection of money from eco-tourism is mainly done by the men.

Before a general meeting of the FUG, the members are informed about the agenda and the issues which are going to be discussed, but they are not consulted about it or asked whether there are other issues they would like to include. Most of the members have no idea or interest in what is in the forest management operation plan. Their concern is rather with the decisions on the use of money that flows from the forest management activities. Many members stated that most of the decisions taken by the executive committee relate to community development investments such as schools, road and embankment construction, installation of water taps, training for income generation activities such as bee keeping, stitching, goat and pig farming, and individual loans for biogas construction. By no means all of these decisions are discussed in the general meeting of the FUG, and it is the executive committee that controls what is on the agenda of these meetings. It is perhaps not surprising then that attendance at these meeting is low, and many people leave the meeting early. Most of the poor members say they do not fully attend the meeting for two reasons: firstly, because the important decisions are made without any consultative meeting beforehand, but also secondly because the meetings are long: they waste one full day's work, meaning that poorer members have to go to bed without food. One poor man from a marginalised group commented that the meeting date is pasted on the executive committee's office board but that he does not participate in any meeting called by the executive committee since it does not solve his livelihood problem, on the contrary, it makes life more difficult. For example, members of the FUG have been prohibited from fishing. Earlier they used to fish in the river for free but after the area was incorporated within the community forest, the executive committee has barred them from this activity, to protect the aesthetic view of the river. As for women: when asked why they did not attend the meetings, most of them responded that they do not like to attend the meeting because they sit at the back and don't hear what is being discussed and even if they put forward some ideas for discussion, their agenda is ignored. The result is that these "weaker groups" are little exposed to new information and knowledge in forest management, a fact which has been noted by other researchers in Nepal. (Neupane, op.cit.).

Distribution of the Forest Products

Power relations are crucial within community forestry because in many user-groups it is the socially dominant individuals who are influential within the management committee, yet it is believed to be the more marginalised members who are more dependent on forests and harvest the majority of the forest resources (Nightingale, op.cit.). All members pay membership fees and collection fees for forest products. In Baghmara BZCF the members are allowed to harvest firewood twice annually, and this is usually done during the big festivals (*Dasain* and *Maghi*). For every 100 kilos of firewood a member has to pay 50 rupees. On the other hand, grass and fodder may be collected throughout the year and there is no fee attached to this activity. In the case study area, it seems that firewood collection is carried out by both better-off and poorer families, although some poor families sell part of their share to middle class and wealthy members. Other studies in Nepal indicate that the better off families may in fact be collecting much more firewood than poorer families (Neupane, op.cit.). However in Baghara some women from poor and marginalised groups commented that they are unable to pay the collection fee as they don't have enough money. A few claim that that the Chief Warden of the Park has instructed committee members to distribute firewood free of cost to the poor members but that the committee has not done this. A number of women of the Musahar tribe say that although after paying the fee they are allowed to go inside the forest to collect firewood like all women members of the FUG, their group is instructed not to collect large branches, while women from more affluent groups collect large branches with impunity. If they are caught with larger branches, then the committee people reprimand them, and tell them they have to pay extra money. This is despite the fact that they do not have sharp sickles and are thus unable to cut as much wood as the high caste women. Their men folk cannot afford the time to collect wood because they have to go to work. Two days of patrolling and other forest work is obligatory for all male members, who in return are allowed to take a load of firewood on those days, but according to informants of the Musahar tribe the amount of firewood allowed is so little that it hardly lasts a few days for a large family.

Although grass and fodder may be collected throughout the year, and no fee is charged, even this does not always result in an equitable distribution. Unlike other groups, the Musahar do not gather fodder from the forest, since they do not possess cattle. Since fodder is, in term of volume, the major non-timber product of the forest, and given their complaints about the way they are hindered in firewood collection and fishing, some Musahar women are beginning to question whether it is worth being a member of the FUG at all.

Yet the Musahar are the most vulnerable group in the whole community and depend more than any other group on natural resources. Evidently, the regulations and system of fees that have been introduced by the FUG are not really conducive to participation by this group, and create asymmetry in the sharing of resource benefits. It seems that even after ten years of operation, the Baghmara FUG is unable to address this problem.

DISTRIBUTION OF OTHER BENEFITS OF FOREST MANAGEMENT

Apart from firewood and fodder, which are direct products, considerable income is derived from the forest from the sale of timber, from the collection fees, from eco-tourism, and from funds from other organisations. For example, in 2006 Baghmara BZCF was awarded the prestigious King Gyanendra Nature Conservation Award, with prize money of 100,000 rupees, by the Royal Nepal Academy of Science and Technology (RONAST), for contributing to sustainable development by promoting eco-tourism and conservation of biodiversity through community forest management. These funds are used to support a variety of community development projects. Many of these are of a general nature and in principle benefit the village as a whole (road improvement, embankments, schools etc), but others are targeted towards individuals, in particular the projects for training in income generation activities. These include bee-keeping, seasonal vegetable farming and animal farming. In addition, financial support is given to individual families for construction of toilets, rice husk stoves and biogas plants, in the form of loans.

These benefits do not reach all families equally. The Musahar women mentioned that they have not received any kind of training, only few are enrolled in adult literacy classes. In any case they do not have sufficient money to start any micro enterprise and cannot raise animals as they do not have land. So although the programmes devised by the executive committee are intended for poor and marginalised women, they are often in practice of little relevance to them. Most of the training sessions and workshops are in fact attended either by the wealthy or the middle class groups. "Weaker groups" are unable to attend as they are day labourers, and their families will go hungry if they miss a day's work (the workshops generally provide a meal for the participants, but the families of these participants of course do not get fed). One woman member of the executive committee explained that they try hard to bring poor and landless people into income generation training but they do not come. Most marginalised people, the poor and particularly poor women indeed leave their houses early in the morning to work as labourers in the road or building construction industry in the city and return home only after dark.

As regards the issuing of loans for the purchase of equipment, particularly for biogas, the "weaker groups" say that they do not benefit at all. The research showed that biogas is mostly installed in wealthy and middle-class houses, which is not surprising as the loan only covers part of the total cost, and only these families are able to pay the extra money needed for the installation. Moreover, it is only the wealthy and middle class that have enough cattle to supply dung for a biogas plant, and can afford to stall-feed them close to the house, which is necessary for transferring the dung to the biogas plant. The poor have fewer (or no) cattle, and lack the space to build stalls close to their houses, and the time to gather fodder for stall feeding. The poor do not take loans for other equipment such as toilets and husk stoves because they do not have any collateral and in any case they often have difficulty paying back the interest.

From this one can conclude that distribution of the benefits of the community forest management effort are not equally distributed within the community. It is not necessarily the case that this mal-distribution is deliberate on the part of the FUG and its executive committee, although the exclusion of the Musahar people does seem to indicate on-going bias. It is more that there is deep-rooted, structural inequality within the village already, which is very difficult to overcome. Indeed it would be very surprising if a single programme like community forest management were able to totally change these economic and social relationships, although recognition of the problems, and efforts to design community forest management procedures which take them better into account, could certainly be improved.

THE FATE OF CARBON FUNDS IN THE FUTURE

If the local community were to be rewarded in financial terms for the carbon saved as a result of their forest management, would principles of equality hold, and would the poorer and less powerful part of the population, and women, benefit at all? The preliminary findings from the case study in Baghmara Buffer Zone Community Forest as regards the current distribution of benefits indicate that particularly as regards financial benefits, it is the richer parts of the population who gain most, even though most of the poorer people (Musahar excepted), and women, get a fair share of the products in terms of fodder and firewood. This outcome is not surprising since it is the men of higher caste and income that get to make the main decisions, despite the idea that the FUGs are supposed to be run on democratic lines. Whether this pattern would be repeated if a greater financial reward is entered into the system through sale of carbon sequestered or deforestation avoided, is not entirely clear. For example, one of the main reasons why the richer families benefit is because they are able to take loans for certain equipment from the community forestry fund; they have the means to match loans and collateral against the repayment. If money for carbon were not handled in the form of loans but (at least in part) distributed to members directly as an annual payment, then this problem should be overcome, and indeed the poor people would stand to earn a welcome, if small, additional income. It remains to be seen whether rules on membership would be tightened to limit membership in some way, if the financial rewards from carbon credits were considerable. At present membership is all inclusive. All this implies is that if equity goals are to be taken seriously, some serious consideration needs to be made regarding how the whole system of rules and procedures for internal payment of carbon services is to be designed, and that particular attention needs to be paid to how the needs and rights of the "weaker groups" will be guaranteed.

SITUATIONS CAN FOREST MANAGEMENT FOR CARBON COMPETE

The sites which were selected for this study are all in places where historically, degradation is the main process by which forest carbon was being lost. These are zones of rather low land value, where there is no obvious competition for alternative land use such as agriculture, because of the terrain, infrastructure such as irrigation, or because of the distance from markets. In such areas, the opportunity costs are low, and a small reward for carbon stock increase or for reduced carbon emissions may represent an attractive financial opportunity. In zones close to cities or in areas of high agricultural potential, there is more likelihood of wholesale land clearance (deforestation), and the value of carbon is probably not sufficient to counteract these processes. The areas studied were all also managed by communities, rather than individual landowners or land users. This means that the unit forest size is in the range of 50 to 600 hectares, an area in which carbon stock can easily be measured and monitored by two or three people in two or three days. This offers considerable economies of scale over individual landholdings, where each individual would have to be trained, or for each of which a special arrangement would have to be made to carry out the stock assessments. It is also the case that forest degradation is related to uncontrolled community land uses (open access behaviour) while deforestation is more likely to be the result of individual land management decisions, as for example in the Amazon frontier where individual settlers (legal and illegal) move in and clear forest for pasture or for cropping, and the West African rain forest belt where timber companies may (legally or illegally) engage in clear felling. Thus there are several reasons why community forest management is particularly well placed as regards crediting for reduced emissions from degradation.

LOCAL TRANSACTION COSTS OF MEASUREMENT

The case studies have also demonstrated the utility of handheld computers with GIS/GPS equipment which make possible accurate mapping of the forest areas and which facilitate the storage of data on carbon stock. This seemingly 'high-tech' approach was found to be very suited to the local conditions, and village people with only a few years of primary education were able to use it after only a day's training (most of them were quite experienced in using mobile phones, which, anno 2006, are common even in the most remote villages). Indeed, local people were quick to recognize the power of such a mapping system and the additional uses to which it could be put (resolution of boundary problems with neighbouring villages etc).

Of course, maintenance of the equipment is another matter, including recharging of the computer batteries (most of the villages were off the grid), scanning into the computer a suitable basemap, and setting out the sampling plots. For these activities it is clear that an NGO or private sector organization with some technical expertise is essential. Given this, and the cost of the computers (about $500, with a similar amount for the software) it is also clear that if the carbon stock change assessments are to be made in a cost effective manner, community forests would have to be clustered into groups, with an NGO or umbrella organization with one set of equipment assisting in perhaps 20 or 30 such forests.

Financing the Carbon

If community forest management is to be employed as a means for reducing emissions from deforestation, and particularly from degradation, in developing countries, then mechanisms to support this will be required, of which financial mechanisms will play a central role. There are issues as regards both international channels for finance and local channels for finance within the countries concerned. Here these are considered from the point of view of community forest management and how such communities could be rewarded for involvement in reducing emissions from deforestation.

International Finance Mechanisms

At the international level, two distinct modes for finance of 'reduced emissions from deforestation' are under discussion, although some combination or hybrid would be possible. The first draws inspiration from the Kyoto flexible mechanisms, and might be placed under the Kyoto Protocol through amendment or in a re-negotiated agreement relating to the second commitment period, which is expected to cover the years 2012-2017. In this approach market mechanisms are central; carbon credits are issued per ton of carbon emission reduced or sequestered, and in principle payment is made ex-post on the basis of this output. The idea is that these can be used directly to meet emission reduction goals. Within this general model of finance tied to carbon credits, there could be two possible ways in which this could be organized: either at a project level, as in the current CDM, with a project specific baseline representing the 'business as usual' rate of deforestation, or at a sectoral level, as in the proposal for 'Compensated Reductions', Party voluntarily accepts a national level target as regards emissions from deforestation, and the baseline is based on national rates of deforestation in the recent past. Either way, any reduction in the observed rate of deforestation compared to the baseline would be translated into tons of carbon, which would have to be verified and certified in some way before they can be sold. The important difference between these two versions is that in the first, the actors on the ground who are responsible for the reductions are directly involved in the deal, as with any CDM; in the second, the marketing deal is with the nation state, and nation state itself would decide how to distribute incentives or payments to encourage the actors on the ground to cooperate in reducing deforestation, or in what other ways the funds generated were to be used.

The second, quite different model for international finance is more in line with a traditional ODA approach to forestry. Financial assistance could be pledged to support efforts to counter deforestation, with a view to reducing emissions but without a direct or quantitative link to the number of tons of carbon saved, and without a direct link to the reduction targets. Such an agreement might fall directly under the UNFCCC rather than the Protocol, or indeed under another international agreement relating to forestry. Funds would be made available by states to support technical assistance and training, forest monitoring and inventory work, and other development activities with a view to helping developing countries implement policies and measures to effectively counter current rates of deforestation and degradation.

There are of course advantages and disadvantages relating to each of these models, which are much under debate at the moment. Many observers feel that ODA funds for forestry have had limited effectiveness as regards reducing deforestation in the past. Moreover they fear that if this model is used, the ODA payments will remain voluntary and not be forthcoming in large enough amounts to really change the current situation. It can be argued that only if payments are linked to legal and obligatory targets (as is the case with carbon reductions under the Kyoto Protocol) will there be sufficient pressure on countries to contribute the funds that would be required, rather than just making token payments. There is also the understanding that a market system will be the most efficient in selecting the most economical carbon mitigation opportunities. On the other hand, the causes of deforestation and degradation are not simple; they result from combinations of many different factors, many of which cannot be tackled directly or individually. A holistic, developmental approach which provides opportunities for alternative livelihoods may be the best way to deal with the problem, but to relate this directly to observable reductions in emissions could be very difficult indeed, given the many drivers and causes at work, and the variations in this in different parts of the world and indeed in different parts of any country. It can be argued therefore that it makes little sense to fund reducing deforestation on the basis of simple carbon output.

Proponents of the market-based approaches take the view that the only means to stimulate real and sufficient investment by countries is to tie this to performance and to binding caps of some sort. The current reduction quotas (average of 5.2% reductions over 1990 emissions) were negotiated before deforestation was considered as a CDM option, and clearly if reducing deforestation were to be admitted as a mitigation option in a Post-Kyoto regime, these caps would have to be re-negotiated, otherwise the market value of carbon would be threatened. Some have proposed that there should be a two target system, one for reductions in fossil fuel emissions, and a separate, but equally binding one, for bio-carbon emissions, including those from deforestation.

The advantage of carbon credits tied to projects is that the savings can be pinpointed easily to particular project activities and investments, as in any CDM arrangement. The major disadvantage and difficulty of including deforestation under the CDM approach is that it is very subject to leakage, through displacement of the deforestation activities to other sites, which is virtually impossible to avoid and very difficult to account for. For that reason a national approach in which the average rate of deforestation over a whole country is measured, rather than individual sites, is much to be preferred (as in, for example, the Compensated Reduction approach). This could, where necessary and sensible, be modified to refer to particular regions within a country. It could in principle also be expanded to cover a multi-nation region (to account for cross-border leakage which is common in many places) although this would make for a more complicated international agreement as regards sharing the credits.

Finance Mechanisms at the National Level

Whichever of the two basic models is eventually selected – a market based, carbon credit system or a system of greatly increased ODA financing to the forestry sector, focusing on reducing deforestation – there remains the question of how such funding is deployed within the country itself. It is noted that many countries have had considerable difficulty in controlling rates of deforestation in the past, not least because of increasing demand for timber products globally, but also because of internal pressures and competition for the use of land. In many cases the economic rent on retaining forest is so much lower than the potential rent from other activities that it is virtually impossible to prevent such shifts. This has of course a lot to do with the fact that the 'true' value of forest (its long term, environmental, intrinsic, and global value) is not reflected in the market system which drives such clearance. Policy mechanisms that can be used to control deforestation and degradation of forest within a country fall into three general categories, in common parlance referred to as 'sticks, carrots and sermons'. 'Sticks' are punitive measures designed to discourage activities leading to loss of forest; they include fines and other punishments for those who infringe laws and regulations designed to protect it. 'Carrots' are positive incentives such as payments for environmental services, or other rewards for not destroying forest. 'Sermons' refers to a wide range of informational activities which in different countries may be referred to as 'raising awareness' or 'education' of local people about the value of forest, or 'motivating' people to reduce their forest-destructive practices, through persuasion. Naturally, forest policy can rest on a combination of carrots, sticks and sermons.

Finance is required for measures in each of these categories, as well as to monitor closely the actual situation as regards deforestation/degradation. Clearly, whatever the package or mix of measures selected, some finance will be needed centrally to pay for the overall management and for activities that need input from the centre, while other finance will need to be distributed, particularly in the case of 'carrots', but also for the implementation of measures of the 'stick' and 'sermons' sort. The appropriate balance will be different in every country. Table gives a sketch of some of the possibilities. Here it is important to consider deforestation separately from degradation, since the two processes may have quite different drivers, and thus may require quite different counter-measures. Apart from other reasons, as already noted degradation often needs to be tackled through an organization at community level, since it affects the common property resources, while deforestation may be more often associated with individual or state land holdings and would need a different organizational approach.

Finance Mechanisms to Local Community Level

Most measures employed by states in the past have not made the distinction between deforestation and degradation and have generally been of the 'stick' or 'sermon' type, but there is currently a movement which is suggesting that 'carrots' might be more effective, at least in some combination with these more traditional methods. These could be targeted at local communities who are engaged in forest management, as well as individual forest land owners in some cases. Particularly for the case of degradation, there is a good case for Payments for Environmental Services (PES) as a tool which can be used at national level, with countries such as Costa Rica and Mexico experimenting with payments to local communities and land owners for water, carbon and biodiversity services. Such an approach would have the advantage of transparency, and thus increase confidence of the international market in the validity of the carbon; moreover, there are undoubtedly many international carbon buyers who would require information on the origin of the carbon, because they have an interest in the knowing that the carbon sequestration is also benefiting the local people, and is not being produced in such a way that they loose their livelihoods, which is a fear that many hold with regard to afforestation and reforestation CDMs.

Such a system could only be employed in areas where communities operate as communities and have the mandate and the ability to organize themselves effectively to manage forest, or alternatively where individuals are legally owners of forest land and have the option of managing it for carbon rather than, or in addition to, other products. This tends to be the case in areas which have a long history of settlement and where population pressure and lack of alternative production potential are driving people to degrade forests to supplement their income. It is much less the case in so-called forest frontier zones where forest is being opened up for the first time: here, deforestation is a greater threat than degradation and the opportunity costs are high. Thus an 'internal system for CDM for avoided deforestation' could only ever form part of a total national approach. Nevertheless, this form of carbon payment to communities for avoidance of degradation could become one very interesting sector of an overall national programme on the lines of 'compensated reductions'.

Another area in which local communities might be directly involved is in monitoring. In most countries, national data on deforestation is poor and unreliable, and data on degradation rates is completely unknown. As the case studies in this booklet show, local communities are well able to make accurate forest inventories themselves, with minimal training, and if these are repeated at intervals to establish rates of change. In a system in which rates of deforestation, and particularly degradation, need to be carefully and reliably monitored so that the state can claim compensation for carbon emission reduction, up to date local level data is going to be essential. It is clear that even if deforestation rates can be established from remote sensing imagery (and this is still in dispute), loss of carbon stocks due to degradation can only be reliably measured at the ground level. Thus there is a necessary role for local monitors, and, as our research shows, this role can easily be taken on by local people with very low levels of education. Their payment for such work would be a necessary part of the transaction costs associated with certifying the carbon credits claimed.

Index